T0214225

SpringerBriefs in History of Science and Technology

The *SpringerBriefs in the History of Science and Technology* series addresses, in the broadest sense, the history of man's empirical and theoretical understanding of Nature and Technology, and the processes and people involved in acquiring this understanding. The series provides a forum for shorter works that escape the traditional book model. SpringerBriefs are typically between 50 and 125 pages in length (max. ca. 50.000 words); between the limit of a journal review article and a conventional book.

Authored by science and technology historians and scientists across physics, chemistry, biology, medicine, mathematics, astronomy, technology and related disciplines, the volumes will comprise:

1. Accounts of the development of scientific ideas at any pertinent stage in history: from the earliest observations of Babylonian Astronomers, through the abstract and practical advances of Classical Antiquity, the scientific revolution of the Age of Reason, to the fast-moving progress seen in modern R&D;

2. Biographies, full or partial, of key thinkers and science and technology pioneers;

3. Historical documents such as letters, manuscripts, or reports, together with annotation and analysis;

4. Works addressing social aspects of science and technology history (the role of institutes and societies, the interaction of science and politics, historical and political epistemology);

5. Works in the emerging field of computational history.

The series is aimed at a wide audience of academic scientists and historians, but many of the volumes will also appeal to general readers interested in the evolution of scientific ideas, in the relation between science and technology, and in the role technology shaped our world.

All proposals will be considered.

Alison Kraft

From Dissent to Diplomacy: The Pugwash Project During the 1960s Cold War

 Springer

Alison Kraft
Max Planck Institute for the History
of Science (MPIWG)
Berlin, Germany

ISSN 2211-4564   ISSN 2211-4572   (electronic)
SpringerBriefs in History of Science and Technology
ISBN 978-3-031-12134-0   ISBN 978-3-031-12135-7   (eBook)
https://doi.org/10.1007/978-3-031-12135-7

Photographs: Photographs reproduced with the kind permission of British Pugwash and the Churchill Archives Center, University of Cambridge. All photographs taken by Jiří Plechatý.

Wiesner diagram: Image reproduced with the kind permission of British Pugwash and the Churchill Archives Center, Churchill College, the University of Cambridge, UK.

This Springer imprint is published by the registered company Springer Nature Switzerland AG
The registered company address is: Gewerbestrasse 11, 6330 Cham, Switzerland

Acknowledgements

This book was made possible by the generous support of the Max Planck Institute for the History of Science (MPIWG) in Berlin, and I would like to extend warm thanks to several colleagues in Dahlem, beginning with Jürgen Renn, Florian Schmalz, Jürgen Kocka, and Carsten Reinhardt. I am indebted to Lindy Divarci for her invaluable expertise and guidance throughout this project. I am very grateful to Matteo Valleriani for the opportunity to present my work in the SpringerBriefs History of Science and Technology series. Warm thanks go to my colleagues in the History of the Max Planck Society project, especially Martina Schlünder, Thomas Steinhauser, and Juan Andre Leon Gomez, for stimulating intellectual exchanges and ongoing support. I thank my MPIWG colleagues Roberto Lalli, Giulia Rispoli, and Jeahwan Hyun with whom, in 2019/2020, I worked on the MPIWG Seminar Series 'Science, Technology and Diplomacy during the Cold War and Beyond: Frameworks, Perspectives, and Challenges.' Warm thanks go to Kristina Schönfeldt, Lina Schwab, and Mona Friedrich both for their collegiality and administrative support. I wish also to thank Nana Citron for excellent help early in the process of preparing the manuscript.

Special thanks go to Carola Sachse: We began working together on the Pugwash organization in 2012, a collaboration that has been intellectually rewarding and an experience from which I have learned a great deal. I am also grateful to Matthew Evangelista and Paul Rubinson for their generosity in sharing their insights on Pugwash over the last few years. Thanks go also to Bernd Greiner and Bettina Greiner at the Berlin Center for Cold War Studies where I held a research fellowship in 2017/2018. During this time, I met Elke Seefried to whom I extend warm thanks for ongoing and insightful discussions. I thank colleagues in the Department of History at the University of Nottingham where I have a long-standing connection.

The book was written largely in the MPIWG library, which is unrivalled as a place to work for scholars working across the spectrum of the history of science. This privilege is something I greatly appreciate, and I wish to thank all the library staff, and Ellen Garske, Ruth Kessentini, and Matthias Schwerdt in particular, for their outstanding research support. I would also like to thank my colleague in the North Wing of the library, Skúli Sigurdsson, for stimulating discussions on the history of science as well as on current events, art, and music.

Much of the research on which this book is based was carried out at the Churchill Archives Center, Churchill College, University of Cambridge, UK, and I extend my thanks to the archive team there for their assistance in the reading room. I am indebted both to this center and to Sally Milne and Andrew Gibson of the British Pugwash Group, for permission to use the photographs and the Wiesner Diagram that appear in the book. I am very grateful to an anonymous referee for useful comments on an earlier version of the manuscript, and extend thanks also to the team at Springer for their help on the production side of the publication process.

I also owe an enormous debt of thanks to my family, to my sisters Sarah and Louise, to Dave, Paul, Dawn, and Matthew, and to my friends: Sue, Jacky, Margaret, Maggie, Janet, Grindl, Sophy, Andreas, Tony, Willi, Tim, Roger, Susanne, Beatrix, Gail, Frances, Caroline, Louise, Ana Maria, Pam and, for the music, McKinley.

Finally, I thank Liz, for all of the laughter and adventures, as well as for the support and companionship without which this book would not have been possible.

Contents

Abbreviations

AAAS	American Association for Arts and Sciences
ABM	Anti-Ballistic Missile
BW	Biological Weapons
CERN	Conseil Européen de la Recherche Nucléaire (European Organization for Nuclear research)
CORIOPAS	Conferences on Research on International Peace and Security
EPG	European Pugwash Group
FAO	Food and Agriculture Organization (UN)
FRS	Fellow of the Royal Society (UK)
GCD	General and Complete Disarmament
ICSU	International Council of Scientific Unions
IAEA	International Atomic Energy Agency
IPRA	International Peace Research Association
ISS	Institute for Strategic Studies (UK)
KGB	Komitet Gosudarstvennoi Bezopasnosti (Committee for State Security, USSR)
LTBT	Limited Test Ban Treaty
MIT	Massachusetts Institute of Technology
MoD	Ministry of Defence (UK)
NATO	North Atlantic Treaty Organization
NGO	Non-Governmental Organization
NPT	Non-Proliferation Treaty
PCSWA	Pugwash Conferences on Science and World Affairs
PSAC	President's Science Advisory Committee (US)
PSGES	Pugwash Study Group on European Security
SADS	Soviet-American Disarmament Study Group
SIPRI	Stockholm International Peace Research Institute
UN	United Nations
VDW	Vereinigung Deutscher Wissenschaftler
WHO	World Health Organization

Chapter 1
Introduction

Abstract Formed in 1957, the Pugwash Conferences on Science and World Affairs (PCSWA) was a unique cross bloc project conceived by natural scientists with the aim of working towards nuclear disarmament. Its paradoxical vision was to pursue that goal by both working with and challenging state actors. Pugwash conferences became sites of East–West dialogue between senior scientific and political elites in the Cold War. This chapter outlines the early development of the PCSWA, its Continuing Committee (its de facto leadership), and the important role of the Secretary General. It highlights the dissenting roots of Pugwash in the West, its narratives about the shared language of science, its demonstrative commitment to political neutrality, and its pioneering role in informal 'Track II' diplomacy. It argues that the development of Pugwash reflected a constant interplay between external political factors and an internal dynamic that, for all the rhetoric about informal modes of working, was deeply hierarchical.

Keywords PCSWA; Pugwash · Russell-Einstein Manifesto · Vienna Declaration · Secretary General · Continuing Committee · Political neutrality · émigré scientists · Joseph Rotblat · Eugene Rabinowitch · Science diplomacy · Vietnam War · The German problem

This book explores key moments of transition and emerging challenges in the early history of a unique organization—the Pugwash Conferences on Science and World Affairs (PCSWA; Pugwash). The PCSWA sought to apply the expertise of natural scientists to problems in world affairs—initially, specifically the nuclear arms race and disarmament—at the height of the early Cold War. Powered by ideas about scientists' social responsibility, brandishing scientific expertise, emphasizing the 'shared language of science,' and claiming political neutrality, Pugwash sought to become a "strong force for peace" (Pugwash 1960).[1]

---

[1] Joseph Rotblat. 1957. Memo for first meeting of the Continuing Committee, 15 December 1957. RTBT 5/2/1/1–15, The papers of Sir Joseph Rotblat, Churchill Archive Center, University of Cambridge, UK. (Henceforth, documents designated by the Churchill catalogue code for the Rotblat collection: RTBT).

© The Author(s), under exclusive license to Springer Nature Switzerland AG 2022   1
A. Kraft, *From Dissent to Diplomacy: The Pugwash Project During the 1960s Cold War*,
SpringerBriefs in History of Science and Technology,
https://doi.org/10.1007/978-3-031-12135-7_1

Writing in 1969, the American physicist and occasional participant at Pugwash conferences Victor Weisskopf exemplified claims made by natural scientists about the special status of the natural sciences, and about a shared language of science:

> Science is a truly human concern; its concepts and its language are the same for all human beings. It transcends any cultural and political boundaries. Scientists understand each other immediately when they talk about their scientific problems, and it is thus easier for them to speak to each other on political or cultural questions and problems about which they may have divergent opinions. The scientific community serves as a bridge across boundaries, as a spearhead of international understanding (Weisskopf 1969, 34).

Such ideas were valuable resources for the founders of the PCSWA, and became part of its identity. This narrative had powerful effects within the organization, not least feeding into its belief in the ability of the "scientific community" to transcend ideological and political divides, some sense of which was reflected by Hermann Bondi (1919–2005):

> One of the continuing great advantages of Pugwash is that one meets people from the other side in whose integrity and intelligence one has every confidence. In this way one can really appreciate the fears, hopes and general motivations on which the policies of other countries are founded (Roxburgh 2007).[2]

Beginning in 1957, Pugwash scientists—at this point, mainly physicists—developed a distinctive approach to conflict moderation centred on regular conferences. This book analyzes a sequence of these conferences in order to shed new light on the history of this organization. It builds upon recent scholarship to generate further insights into how its scientists were able to work across the blocs to develop new approaches to disarmament, to forge new modes of informal diplomacy, and to expand the scope of its work to address emerging issues within the realm of global security amid the changing contours of the 1960s Cold War (Kraft and Sachse 2020b). One focus lies with the Continuing Committee. Formed in December 1957, this functioned as the de facto leadership of the international PCSWA and throughout the first five years, its membership comprised exclusively scientists from the US, USSR and UK. Its founding members were the Americans, Hiram Bentley Glass (1906–2005), Harrison Brown (1917–1986), and Eugene Rabinowitch (1901–1973); the Soviet scientists Evgenii Federov (1910–1981), Dimitrii Skobel'tsyn (1892–1990) and Aleksandr Topchiev (1907–1962); and the Britons, Neville P. Mott (1905–1996), Rudolf Peierls (1907–1995), and Cecil F. Powell (1903–1969), together with the Secretary General, Joseph Rotblat (1908–2005).

In the West, the Pugwash project was initiated and led in its early years by a small group of mainly British political outsiders with a proud dissenting tradition, most notably perhaps Bertrand Russell. Described in 1998 by his Pugwash colleague, the Polish émigré physicist Joseph Rotblat, as "the great dissident of this century," Russell together with the French physicist Frédéric Joliot-Curie were the architects of the Russell-Einstein Manifesto of 1955—long regarded as the inspiration for

---

[2] Hermann Bondi, Thoughts on Pugwash, 28 April 1967. Response to Joseph Rotblat's memorandum on the future of Pugwash, April 1967. RTBT 5/3/1/19.

Fig. 1.1 Joseph Rotblat (1908–2005). Secretary General of PCSWA, 1957–1973. Pictured during the fifth Pugwash Symposium held in Marienbad, 19–24 May 1969. RTBT 5/2/2/5 (17)

the PCSWA (Roberts 2020). They went on to organize the inaugural conference in July 1957—strongly assisted by British colleagues Eric H. S. Burhop (1911– 1980), Cecil F. Powell and Joseph Rotblat, all physicists who shared a track record of challenging the British government on nuclear matters and policy (Kraft 2018). They took this tradition of dissidence and solidarity forward into the international PCSWA, as Rotblat put it:

> It has been a desideratum of the Pugwash Movement to be progressive, to foster new ways of thinking, to encourage pioneering ideas. Occasionally, this may bring us into conflict with the establishment; it may make us nonconformists, radicals, dissidents. Dissidence can be said to be part of our ethical code (Rotblat 1998).

By the early 1960s, Russell's role within Pugwash was waning; by contrast, Rotblat's influence was rising, an influence explained by his role as Secretary General (Braun et al. 2007; Hinde and Finney 2007; Kraft and Sachse 2020b) (Fig. 1.1).

As an avowedly cross-bloc project working for peace, conceived and built by maverick western 'outsiders', the PCSWA was from the outset viewed with suspicion if not outright hostility in London and Washington. This was also linked to the fundamental asymmetry embedded within the organization: namely, the very different relations pertaining between Pugwash scientists and their national governments in East and West. Within the Soviet Union, Pugwash scientists were selected by the state, via the national Academy of Sciences, for their political reliability. They typically enjoyed close relations with the Kremlin which, from the outset, was supportive of Pugwash. By contrast, in the West, leading Pugwash scientists, including Rotblat, were seen within government as politically unreliable. Pugwash presented Whitehall and the White House with a dilemma: not only were they unable to control the western scientists building Pugwash, but Rotblat and others in the founding cohort, notably the St. Petersburg born, American émigré biophysicist and cofounder/editor of the *Bulletin of the Atomic Scientists,* Eugene Rabinowitch, were now—and on their own terms—building relations with Soviet scientists handpicked by and close to the Kremlin (Slaney 2012). Paradoxically, if this dynamic filled the western political establishment with unease, it was exactly this proximity to Moscow that in the early 1960s began to register differently in senior western political circles. This set the scene for a shift in which, at a particular juncture in the Cold War, it became possible for the scientists of Pugwash to develop a new mode of techno-political communication that translated political problems into science talk and, in so doing, took the organization into the realm of informal diplomacy.

The Pugwash leadership had always to carefully manage relations with the US and the Soviet Union whilst also striking a balance between different constituencies and interests, especially those of East and West, within the organization. Here, the Secretary General was of pivotal importance. In pioneering this role, Joseph Rotblat worked first and foremost to establish the Secretary General as an impartial actor, within Pugwash and beyond. This neutrality was the basis of the authority of this office and, in this period, underpinned the enormous respect accorded Rotblat by his Pugwash colleagues. In short, a politically neutral Secretary General was crucial to how the international Pugwash organization functioned, providing a vital internal mechanism for balancing different interests within the organization—which extended to a role in adjudicating when internal conflicts arose, for example, over the Vietnam War. Meanwhile, an impartial Secretary General was also crucial to external perceptions of Pugwash: it sent a signal to Washington and Moscow that the organization had an internal mechanism for ensuring evenhandedness in its relations with and treatment of East and West. Against the odds, Joseph Rotblat made this work. In this formative period, his developing reputation for impartiality, and as a man of integrity, was an asset in the narrative about the political neutrality of Pugwash on the international stage. Both attributes were key to his pivotal role in the organization's developing involvement within the realm of informal diplomacy, during and beyond the conferences.

This book casts fresh light on the transformation of the PCSWA from its founding form in the West within a network of dissenters to an organization respected—even trusted—by governments on both sides of the Cold War divide. It shows how changing perceptions of Pugwash in the West were decisive in this shift and how, in turn, this made it possible for the organization and its scientists to become pioneers in Track II diplomacy (Davidson and Montville 1981–1982; Jones 2015; Pietrobon 2016). The book also highlights the difficulties faced by Pugwash as it struggled to meet the new challenges of the mid-1960s Cold War. It identifies and investigates two overarching trends that defined this period: on the one hand, the way in which the Vietnam War caused internal tensions and made for increasingly difficult relations with the Lyndon B. Johnson administration in Washington, and on the other hand, the growing influence of European scientists within the organization who were determined to address issues relating to European security. Both trends, which emerged at the same time, posed serious difficulties for the Pugwash leadership.

Earlier histories of the PCSWA chronicled the organization from an insider viewpoint, portraying its scientists as honest and reliable brokers working for peace, and celebrating its achievements and legacies (Rotblat 1967; Schwartz 1967; Brown 2012). Since the 1990s, scholarly analyses have been undertaken from new perspectives. The approach of transnational history with its focus on non-state actors has richly illuminated the transnational dimensions of the PCSWA, most prominently in Matthew Evangelista's influential 1999 book, *Unarmed Forces*, which focuses on the Soviet case and remains the only in-depth, country-based book-length analysis of Pugwash (Evangelista 1999; Turchetti et al. 2012).

A number of studies have examined the work of the Pugwash organization in relation to particular arms control treaties, most notably Bernd Kubbig's analysis of the Anti-Ballistic Missile Treaty, Paul Rubinson's study of the Limited Test Ban Treaty (LTBT) and Kai Henrik Barth, who emphasized its importance for both the LTBT and the 1968 Non-Proliferation Treaty (NPT), describing Pugwash as "the most important transnational effort of scientists in the Khrushchev/Brezhnev era" (Adler 1992; Kubbig 1996; Barth 2006; Rubinson 2011). Pugwash has also featured briefly in recent literature examining scientists' responses to the ethical dilemmas posed by nuclear weapons and Cold War ideology (Bridger 2015; Wolfe 2018). Drawing on unpublished as well as published sources produced by the organization, more recent scholarship has examined more closely the development and internal dynamics of the PCSWA, internationally and within different national contexts (Kraft et al. 2018; Kraft and Sachse 2020a; Lüscher 2021). This work has examined critically the organization's central *raison d'etre* and the narrative expounded by the leadership about the special mission of natural scientists to tackle the dangers of nuclear armaments post-Hiroshima. This was made possible, so the narrative went, by the special competence of natural scientists to remain impartial and to suspend national and bloc allegiances in order to communicate objectively across the otherwise intractable political and ideological divides that defined the Cold War. The new analyses of the PCSWA have highlighted its distinctive practices, not least its informal *modus operandi,* a strict code of confidentiality, including adherence to Chatham House rules, namely a "gentleman's agreement that participants do not, during or after the conference, quote other participants' views with their names," and annual gatherings organized on an invitation only basis.[3] The Pugwash conferences replicated the forms and rituals of international scientific conferences but harnessed them to very different purposes, namely working towards disarmament and for peace (Kraft and Sachse 2020b). This work has also drawn attention to the different character of national Pugwash groups, the different pressures operating on national groups in East and West, and their role within the international organization.

For all the insights yielded by this scholarship, a number of questions remain unanswered in relation to the inner workings of the international Pugwash organization and its external relationships and influence, not least with state actors. With regard

---

[3] Harrison Brown. Notes on formal opening session. Seventh PCSWA, Stowe, Vermont, USA. 5 September 1961. RTBT 5/2/1/7 (13).

to developments within the organization, there is agreement in the literature hitherto about the enormous influence which Joseph Rotblat wielded within Pugwash. However, there is more to explore both about the development of his strategy for the organization and about how and why he was able to exercise and maintain his influence within it. Here, it is important to keep in view the paradoxical vision that motivated and sustained Rotblat and other early leading figures within Pugwash, including Eugene Rabinowitch: they sought to mobilize their unique international resource of scientific expertise in order to reach into government circles and to work with governments whilst at the same time challenging those governments on their nuclear weapons policy and their adherence to the logic that was driving the arms race.

This vision and the precarious balancing act it entailed had several consequences for the internal development of the organization. It helped fix the assumption that the strength of the organization lay in mobilizing an elite of natural scientists—rather than scholars from a wider range of disciplines—as the most credible bearers of objective scientific expertise. The book explores the centrality of natural science expertise to the Pugwash project in terms of both its importance as the bedrock of its claims to political neutrality and as the prism through which the leadership viewed and was willing to engage with pressing political issues of the day. However, the science and technology-based rationale for engaging with political issues came under pressure as, in the mid-1960s Cold War, political problems moved increasingly to the fore. This book reveals a number of challenges to Rotblat and the Continuing Committee that emerged in this period from within the organization. It explores the extent to which the leadership was open to new departures initiated from other quarters within the organization, and how the leadership dealt with the tensions that could arise between its internationalist ethos and national-specific agendas. It examines too the way in which private conversations amongst the leadership created the dynamic within Pugwash of an 'inner circle' that guided the official and public presentation of the role and goals of the organization.

In relation to external perceptions of Pugwash and its influence with state actors, research hitherto has established that Pugwash was initially regarded with suspicion by western governments wary of Soviet 'front' organizations and that it developed a role as a respected interlocutor (Kraft 2020; Kraft and Sachse 2020b). Drawing on untapped primary sources, this book investigates further this transition, analyzing its dynamics and the nature of the changes that made it possible. It asks how Pugwash was able to overcome suspicions of it in Washington and London, and why the organization came to appear to Western governments as potentially relevant and useful. It explores the development and basis of its distinctive informal style of working, and the effects of this within the organization. For example, the book examines its importance for the process by which Pugwash became both a forum for and an active agent in 'back-channel' or Track II diplomacy of various kinds, including beyond the conference setting. Indeed, the book argues that Pugwash and its scientists can be considered as pioneers of Track II diplomacy. In seeking better understanding of this crucial dimension of the Pugwash organization, this book casts fresh light on the principal role of the Continuing Committee in conceiving and carrying out 'back-channel' activities. It also draws new attention to the role of and work carried out within Working Groups at the annual conferences. To date, this innovative aspect of the conferences has been overlooked. The Working Groups were key sites of cross bloc exchange and Track II activities, important both in fashioning the distinctive form of technopolitical dialogue developed within Pugwash and as active sites of research into disarmament but also other Cold War problems. In short, these groups became a valuable asset that helped to draw those within powerful political and nuclear policy-making circles, especially in the West, to the annual conferences. As the book also makes clear, Working Groups could also serve as engines for change within the organization.

The Pugwash organization created a new space at the intersection between science, politics, (nuclear) policy-making and diplomacy. This raises questions about its connections to 'science diplomacy,' a concept which has recently come to the fore as a framework for analyzing this intersection. The relationship between science and diplomacy has long been a subject of historical research, including scholarship that has emphasized the use of science as an instrument of 'soft power' in international relations (Nye 1990, 2004; Beatty 1993; Hamblin 2000; Voorhees 2002; Doel and Harper 2006; Krige and Barth 2006; Miller 2006; Heyck and Kaiser 2010; Krige 2016; Turchetti 2018; Whitesides 2019). Another historiographical strand in the literature on science and diplomacy has explored the international role of scientists in conflict resolution (De Cerreno and Kenyan 1998; Salomon 2001). The initial impetus for the contemporary conceptualization of 'science diplomacy' came from within American science policy circles in conversations about the use of science—and scientists—to realize foreign policy goals (Federoff 2009; Turekian 2012; Davis and Patman 2015). While this idea is far from new, the alignments and goals envisaged in this formulation of this relationship reflect the distinctive set of political imperatives operating in the post-Cold War, globalized world and the changing power relations across North–South and East–West divides in the twenty-first century (Skolnikoff 1993). An influential model of 'science diplomacy' was formulated at a meeting of senior scientists at the Royal Society of London in 2009 (The Royal Society of London 2010). Here, 'science diplomacy' was defined as having three dimensions: the use of scientific advice to inform foreign policy objectives ('science in diplomacy'); activities/measures geared to facilitating international science cooperation ('diplomacy for science'), and the use of scientific cooperation to improve relations between countries ('science for diplomacy'). Since then, this model has been analyzed and refined in both policy circles and within a growing body of historical scholarship about relations between science and diplomacy. (Shields 2016; Rentetzi 2018; Turekian 2018; Milam et al. 2020; Turchetti et al. 2020; Adamson and Lalli 2021; Ito and Rentetzi 2021; Kunkel 2021).

The report from the 2009 meeting of the Royal Society cited Pugwash in a section outlining historical antecedents of 'science diplomacy,' highlighting its capacity to "explore alternative approaches to arms control and tension reduction with a combination of candor, continuity, and flexibility seldom attained in official East–West and North–South discussions and negotiations" (The Royal Society of London 2010, 1). This book takes a different tack. Whilst acknowledging the usefulness of the concept of 'science diplomacy' as a means to highlight and explore the intersection between science, politics and diplomacy, it asks whether the PCSWA can be assigned a position as straightforwardly part of that history. As a scientist-led project focused on working for disarmament and peace that was seeking to challenge government policies that defined and sustained the Cold War, Pugwash was out of step with official foreign policy strategies and priorities. As a result, in the West, relations between the PCSWA and its scientists, and governments were initially ambivalent. Although the relationship changed, the conscious efforts of the western Pugwash leadership to maintain some distance from western governments while at the same time trying to work with and influence them, casts doubt on the organization's fit within the framework of 'science diplomacy' as defined in the policy literature. That said, the activities of Pugwash and its scientists as a back channel for unofficial, Track II diplomacy arguably locate it firmly within the realm of science diplomacy more broadly construed which, for example, takes account of scientists acting autonomously on the international stage, in ways and for goals that they themselves have defined.

This book focuses on the history of the PCSWA during a particular phase of the Cold War in the 1960s. With this focus, it picks up on insights from Cold War historiography and the nature of the conflict during those years in order to tackle important and hitherto largely unanswered questions about how shifts in the external geopolitical context shaped the external strategy and the internal dynamics of the organization (Leffler and Westad 2010; Immerman and Goedde 2013). If the superpower rivalry and the nuclear arms race had defined the new world order of the early Cold War, during the 1960s the conflict entered a new phase, shaped by moves towards superpower détente, profound changes within the alliance system of both the US and the Soviet Union, and the emergence of regional powers and regional conflicts not least in the so-called 'developing world'—the countries of which were assuming new 'security' importance as they became proxy sites for the superpower rivalry (Westad 2017). The bipolar model of the Cold War which emphasized the superpower dynamic was giving way to a conflict conceived as multipolar or polycentric and global. This shift is reflected in the changing historiography of the conflict as historians have since the late 1990s paid increased attention to the Cold War as experienced in the countries of the Global South (Geyer and Bright 1995; Connelly 2000; Smith 2000; Westad 2007; McMahon 2013; Mooney and Lanza 2013; von Bressendorf and Seefried 2017).

Two particular issues arising from Cold War developments in the 1960s posed serious challenges for the Pugwash leadership. This book examines in depth for the first time the impact of the Vietnam War on the PCSWA and how the Pugwash leadership handled this conflict. Vietnam created deep divisions and unprecedented tensions within the organization, and soured already uneasy relations with the Lyndon B. Johnson administration in Washington (Rubinson 2020). But in 1967 Vietnam also provided the context at a particular moment for a remarkable behind-the-scenes attempt at conflict moderation by Pugwash scientists. Secondly, the book also explores the organization's venture into the overtly political terrain of European security, which encompassed the 'German problem' (Kraft 2020). This ground-breaking step took the form of a new Pugwash Study Group that was notable in two respects.

Firstly, this was a project driven by scientists from Eastern and Western Europe, and secondly, it brought into Pugwash scholars from disciplines other than the natural sciences, most prominently from economics, law and the social sciences. This new departure raised difficult questions for the natural scientists within the Pugwash leadership and had implications for the profile of Pugwash on the international stage. As the book shows, the challenges of dealing with the Vietnam War and European security revealed both the strengths and the weaknesses of the Pugwash leadership, whilst exposing and testing the limits of the organization's capacity to engage with complex political issues somewhat removed from matters of science and technology.

Amid the changing contours of the Cold War, the leadership was aware that the PCSWA needed to deploy, and potentially extend, its scientific and technical expertise to address the very different challenges of the mid-1960s. The book asks how the leadership responded to these new external pressures and how it dealt with the risk that the diversification of its expertise base posed to the reputation of Pugwash as an organization of natural scientists valued for its scientific and technical expertise. All these challenges came to bear upon the leadership as other institutional actors were arriving on the international stage with a remit similar to that of the PCSWA, namely dealing with disarmament and security issues on a cross-bloc basis. As the book shows, all of this combined to create by 1967 a growing crisis of identity and purpose within the PCSWA.

Focusing on the period between 1960 and 1968, the book sets out to test an overall hypothesis that the development of the PCSWA reflected a constant interplay between external drivers of change that were largely political and beyond its control, and an internal dynamic that, for all the rhetoric of informality, was in practice deeply hierarchical. More specifically, this hypothesis posits that the Pugwash leadership used the informal modus operandi that it had carefully fashioned and steadfastly protected to create and maintain this hierarchy. Power remained concentrated in the hands of a small elite—the so-called Continuing Committee—that formed the apex of the international organization and exercised close control over its constituent parts. This Committee—dominated in this period by natural scientists, especially physicists, form the US, USSR and the UK—enjoyed a great deal of autonomy and enormous discretionary powers within the PCSWA. In this way, the Committee, on which the Secretary General held the only permanent seat, was able to closely control the development of Pugwash, most obviously, the focus of its agenda and the scope of its activities (both at the conferences and beyond). When the changing geopolitical contours of the mid-1960s Cold War fueled calls from some for a diversification of the Pugwash agenda, this Committee moved to ensure that the organization retained its focus on what it perceived to be the single greatest threat to world peace, namely: nuclear weapons, specifically the dangers of nuclear proliferation and of a thermonuclear war involving the superpowers.

To explore this hypothesis, the book sets out to capture through a detailed analysis of hitherto untapped primary sources the character and dynamics of a sequence of Pugwash conferences in the period under study. This fine-grained analysis enables the identification and investigation of continuities and discontinuities, moments of synergy, tension, rupture and of transition within and between these conferences.

Comparisons between the conferences reveal, for example, high points of real dialogue and low points of Cold War grandstanding. Archival sources relating to the conferences include the *Proceedings* which were produced internally and distributed to senior government circles, with copies sent also to the participants at each conference. These volumes could—at the discretion of the Continuing Committee—include the reports of Working Groups. The minutes of Continuing Committee meetings also constitute a valuable resource. These were carefully crafted accounts of what took place and they reveal, in their brevity and silences, the sensitivities of these scientists to the confidential character of these cross-bloc meetings, the fragility of the Pugwash project, and the polarized Cold War context in which they were operating. It is clear that more went on in these meetings than is reflected in the official internal records. Collectively, these sources comprise an official narrative about the conferences and developments within the organization. Other key sources include the records of the European Pugwash Group (1959–1964) and the Pugwash Study Group on European Security (1965–1968) which shed light onto activities in other parts of the expanding Pugwash organization and which involved European scientists positioned differently within its hierarchical structure.

As a counterpoint to the official records, private correspondence between Pugwash scientists provides a rich source of alternative insights. Private correspondence was key to the formation, maintenance and expansion of the scientific networks that in this period defined the Pugwash organization: senior figures routinely kept in touch by letter on a year-round basis. Within this confidential circuitry of letters traces are found otherwise unrecorded news and views of and from colleagues, of side meetings at the meetings of the Continuing Committee, or off-the-record exchanges at conferences. It is here that the tone and demeanor of fractious discussions at conferences were hinted at or drew comment. These sources make visible hitherto unknown inner workings of Pugwash, and tell a more complex story about how it worked across divides, how it functioned in practice, how its scientists navigated the intersection between science and politics, and how they learned to operate in the external political realm. These letters also cast light on the different personalities guiding Pugwash, and on the friendships and networks that flourished within the organization. They reveal the mentalities of this cosmopolitan elite. Notably, several key Pugwash figures in the West came from an émigré background and shared a Jewish heritage. It seems likely that such similarities in background strengthened further the intellectual and cultural affinities that came from being natural scientists. It may also be the case that their particular émigré backgrounds, mainly from Russia and Poland, helped in building relations with their colleagues from the Eastern bloc.

Working together, Pugwash scientists broke new ground at the intersection between science, politics and diplomacy in the 1960s Cold War. In tackling this intersection, the book takes its cue from recent literature that has emphasized the transnational analytical framework as a means to investigate science and scientists in the global Cold War. (Phalkey 2013; Kemper 2016; McNeill and Unger 2010; Van Dongen 2016).

1.1 Outline of the Book and Background Context

This book focuses on a key phase of Pugwash history starting with the sixth conference held in Moscow in 1960. To understand where the organization stood at this point, it is important to gain a sense of the earlier conferences. The first five conferences took place between 1957 and 1959 and were held either in Canada or in Austria. The Canadian venue reflected the financial reliance of the organization in its early years on the American business magnate Cyrus Eaton, who made his summer home in Pugwash, Nova Scotia, available for two of the early conferences. But Eaton's outspoken calls to build relations with the Soviet Union, motivated not least by his business ambitions and his friendship with Khrushchev brought the fledgling PCSWA into disrepute, leading the Continuing Committee to sever ties with him (Sachse 2020). Austria hosted the third and fourth conferences, held in Kitzbühel, Vienna and in Baden in 1958 and 1959 respectively, due largely to the efforts of the physicist Hans Thirring. Through skillful diplomacy, including playing to the Austrian government's post-war strategy of neutrality and its desire to see Vienna become an international East–West gateway, Thirring secured political and financial support for the PCSWA and these conferences (Fengler 2020).

The third conference concluded in Vienna with a large public ceremony where the Pugwash leadership set out its agenda more fully in a statement that came to be known as The Vienna Declaration (1958). Crafted largely by Eugene Rabinowitch in close cooperation with Joseph Rotblat, the Declaration reflected their shared views about the scope and priorities of the Pugwash project. It transformed the founding vision of the PCSWA into a set of practical aims organized into seven areas of work. The Declaration became a central tenet of the PCSWA, and for many it also embodied the Pugwash 'credo' or 'spirit,' that emphasized collegiality and a willingness to listen to different viewpoints. From the outset, much greater importance was ascribed to the Vienna Declaration in the Soviet Union and the Eastern bloc, where it was strongly endorsed by governments and widely publicized. By contrast, in the West, despite the best efforts of the western Pugwash leadership, it was largely overlooked. In Vienna, the Continuing Committee also called for the formation of national groups as a means to extend the reach of the new organization around the world: by 1971, thirty groups had been created (Rotblat 1972; Kraft and Sachse 2020b).

In the course of these early conferences, the American, Soviet and British scientists on the Continuing Committee were getting to know each other. This process included working out if and how they could work together as, individually and collectively, they sought to find their feet in the uncertain terrain between science and politics in the Cold War. The Committee met two or three times each year, and during these meetings this small circle of scientists formed a fragile but nevertheless enduring cross bloc network as together they guided the development of the PCSWA. Notably, between 1957 and September 1963, this Committee remained an exclusively America, Soviet and British affair. Relations within it were constantly tested. Each scientist had to learn the essential skill of simultaneously looking both East and West, whilst also managing relations with colleagues on the Committee and with the political establishment in his particular national context. In effect, this Committee constituted a bold Cold War experiment in international cooperation as its members probed and tested the possibilities for building goodwill and even a degree of trust across the bloc divide. Emerging in this process was a distinctive form of informal diplomacy between its members. During this period, American and Soviet scientists dominated numerically at the conferences, with the carefully balanced delegations of the superpowers far outnumbering both British and European participation. For example, scientists from the US and Soviet Union together accounted for twenty-one of the twenty-five scientists in Baden in 1959 and seventeen of the twenty-six delegates at the fifth conference in Nova Scotia later that year. This reflected the reality—an early sign of the pragmatism that would become one hallmark of the PCSWA—that, for now, both the Committee and the conferences were geared primarily towards opening up a channel for dialogue between the nuclear superpowers: this was the founding raison d'etre of the organization and its first priority. The Committee recognized that success here would create a platform on which the PCSWA could be built.

This book begins with the sixth conference, held over nine days in Moscow in November and December 1960. The second chapter highlights the depth and scope of the discussions on disarmament here and shows why this conference was a key turning point for the organization. The third chapter focuses on the seventh and eighth PCSWA, both held in Stowe, Vermont, in September 1961. These two conferences became an important test of Pugwash in that they took place in a period of acute tensions between the superpowers. Chapter 4 moves outside the conferences, focusing on three targeted interventions by Pugwash scientists in 'back-channel' diplomacy. These were, firstly, an attempt to de-escalate tensions between Moscow and Washington during the Cuban missile crisis in October 1962; secondly, a technical contribution by American, Russian and British scientists in March 1963 to the process that led to the Limited Test Ban Treaty in August that year; and, thirdly, a mission in shuttle diplomacy between Hanoi and Washington conceived and carried out by Pugwash scientists in June and July 1967 that sought to help bring an end to the war in Vietnam (Salomon 2001). This took place in the context of deep divisions along bloc lines over Vietnam within Pugwash which turned the conferences in Venice in 1965 and Sopot in 1966 into Cold War battlegrounds and threatened to tear Pugwash apart. This fed into a deepening sense of crisis within senior Pugwash circles in 1967 as the leadership also faced serious challenges on other fronts. Prominent here were tensions with European colleagues who were using Pugwash as a forum for East–West discussions on the 'German problem.' This European dimension is at the center of the fifth chapter, which tracks mounting pressure to include European security on the Pugwash agenda which, via steps taken at the Dubrovnik and Karlovy Vary conferences in 1963 and 1964 respectively, led to the formation in 1965 of the Pugwash Study Group on European Security (PSGES). It then explores how and why the Continuing Committee came soon to view the PSGES as problematic, and examines the steps it took during 1967–1968 to disband the group. This chapter concludes by analyzing the crisis engulfing Pugwash in this period about

Evangelista, Matthew. 1999. *Unarmed forces. The transnational movement to end the Cold War.* Ithaca: Cornell University Press.

Federoff, Nina V. 2009. Science diplomacy in the 21st century. *Cell* 136: 9–11.

Fengler, Silke. 2020. "Salonbolschewiken:" Pugwash in Austria, 1955–1965. In *Science, (anti-)communism and diplomacy: The Pugwash Conferences on Science and World Affairs in the early Cold War*, eds. Alison Kraft, and Carola Sachse, 221–258. Leiden: Brill.

Geyer, Michael, and Charles Bright. 1995. World history in a global age. *American Historical Review* 100 (4): 1034–1060.

Hamblin, Jacob Darwin. 2000. Visions of international scientific cooperation: The case of oceanic science, 1920–1955. *Minerva* 38 (4): 393–423.

Heyck, Hunter, and David Kaiser. 2010. Focus. New perspectives on science and the Cold War: Introduction. *Isis* 101 (2): 362–366.

Hinde, Robert A., and John L. Finney. 2007. Sir Joseph Rotblat. *Biographical Memoirs of the Royal Society* 53: 309–326.

Immerman, Richard H., and Petra Goedde, eds. 2013. *The Oxford Handbook of the Cold War.* Oxford: Oxford University Press.

Ito, Kenji, and Maria Rentetzi. 2021. The coproduction of nuclear science and diplomacy: Towards a transnational understanding of nuclear things. *History and Technology* 37 (1): 4–20.

Jones, Peter L. 2015. *Track two diplomacy in theory and practice.* Stanford: Stanford University Press.

Kemper, Claudia. 2016. *Medizin gegen den Kalten Krieg: Ärzte in der anti-atomaren Friedensbewegung der 1980er Jahre.* Göttingen: Wallstein.

Kraft, Alison. 2018. Dissenting scientists in early Cold War Britain. The "fallout" controversy and the origins of Pugwash, 1954–1957. *Journal of Cold War Studies* 20 (1): 58–100.

Kraft, Alison. 2020. Confronting the German problem: Pugwash in West and East Germany, 1957–1964. In *Science, (anti-)communism and diplomacy: The Pugwash Conferences on Science and World Affairs in the early Cold War,* eds. Alison Kraft, and Carola Sachse, 286–323. Leiden: Brill.

Kraft, Alison, and Carola Sachse, eds. 2020a. *Science, (anti-)communism and diplomacy: The Pugwash Conferences on Science and World Affairs in the early Cold War.* Leiden: Brill.

Kraft, Alison, and Carola Sachse. 2020b. Introduction. The Pugwash Conferences on Science and World Affairs: Vision, rhetoric, realities. In *Science, (anti-)communism and diplomacy: The Pugwash Conferences on Science and World Affairs in the early Cold War,* eds. Alison Kraft, and Carola Sachse, 1–39. Leiden: Brill.

Kraft, Alison, Holger Nehring, and Carola Sachse. 2018. Introduction to special issue: The Pugwash Conferences and the global Cold War. Scientists, transnational networks, and the complexity of nuclear histories. *Journal of Cold War Studies* 20 (1): 4–30.

Krige, John. 2016. *Sharing knowledge, shaping Europe. US technological collaboration and nonproliferation.* Cambridge, MA: Harvard University Press.

Krige, John, and Kai-Henrik Barth, eds. 2006. Introduction. Science, technology, and international affairs. *Osiris* 21 (1): 1–21.

Kubbig, Bernd W. 1996. *Communicators in the Cold War: The Pugwash Conferences, the U.S.-Soviet study group and the ABM treaty. Natural scientists as political actors: Historical successes and lessons for the future.* PRIF Reports No. 44. Frankfurt am Main: PRIF.

Kunkel, Sönke. 2021. Science diplomacy in the twentieth century: Introduction. *Journal of Contemporary History* 56 (3): 473–484.

Leffler, Melvyn P., and Odd Arne Westad, eds. 2010. *The Cambridge history of the Cold War,* 3rd ed. Cambridge: Cambridge University Press.

Lüscher, Fabian. 2021. *Nuklearer Internationalismus in der Sowjetunion. Geteiltes Wissen in einer geteilten Welt 1945–1973.* Böhlau: Wien.

McMahon, Robert J. 2013. *The Cold War in the Third World.* Oxford/New York: Oxford University Press.

the aims, scope and purpose of the organization as it also faced competition fro other organizations tackling disarmament, and struggled to adjust to the changir geopolitical contours of the mid-1960s Cold War. Matters came to a head at th seventeenth conference held in Ronneby, Sweden in September 1967, when the cris was resolved in ways that reasserted two Pugwash traditions: its primary focus c science and technology and its prioritization of nuclear disarmament. That said, thre sub-themes were specified as part of its agenda—promoting international scientifi cooperation, strengthening engagement with the so-called 'developing world' an fostering a sense of social responsibility amongst scientists. A period of crisis fc Pugwash therefore concluded in ways that reflected the views of the Continuin Committee and reasserted its authority—though leaving open questions about th organization's future role and its distinctive contribution to 'world affairs.'

References

1960. *Pugwash: Its history and aims.* London. Pamphlet: Bestand 456, Akten 492, Bundesarchiv Koblenz, Germany.

Adamson, Matthew, and Roberto Lalli, eds. 2021. Special issue: Global perspectives on scienc diplomacy. *Centaurus* 63 (1): 1–231. https://doi.org/10.1111/1600-0498.12315

Adler, Emmanuel. 1992. The emergence of cooperation: National epistemic communities and th international evolution of the idea of nuclear arms control. *International Organization* 46 (1) 101–145.

Barth, Kai-Henrik. 2006. Catalysts of change: Scientists as transnational arms control advocates in the 1980s. *Osiris* 21 (1): 182–206.

Beatty, John. 1993. Scientific collaboration, internationalism, and diplomacy: The case of the Atomic Bomb Casualty Commission. *Journal of the History of Biology* 26 (2): 205–231.

Braun, Rainer, Robert Hinde, David Krieger, Harold Kroto, and Sally Milne. 2007. *Joseph Rotblat: Visionary for peace.* Weinheim: Wiley-VCH.

Bresselau von Bressensdorf, Agnes, and Elke Seefried. 2017. Introduction: West Germany and the Global South in the Cold War Era. In *West Germany, The Global South and The Cold War, German Year Book of Contemporary History/Vierteljahrshefte für Zeitgeschichte*, vol. 2, eds. A. Bresselau von Bressensdorf, E. Seefried, and C.F. Ostermann, 7–24. Oldenbourg: De Gruyter.

Bridger, Sarah. 2015. *Scientists at War. The ethics of Cold War weapons research.* Cambridge, MA: Harvard University Press.

Brown, Andrew. 2012. *Keeper of the nuclear conscience: The life and work of Joseph Rotblat.* Oxford: Oxford University Press.

Connelly, Matthew. 2000. Taking off the Cold War lens: Visions of north-south conflict during the Algerian war for independence. *American Historical Review* 105 (3): 739–769.

Davidson, William D., and Joseph V. Montville. 1981–1982. Foreign policy according to Freud. *Foreign Policy* 45: 144–157.

Davis, Lloyd S., and Robert G. Patman. 2015. *New day or false dawn?* Singapore: World Scientific Publishing.

De Cerreno, Allison L.C., and Alexander Kenyan, eds. 1998. *Scientific cooperation, state conflict: The roles of scientists in mitigating International discord.* Annals of the New York Academy of Sciences, vol. 866. New York: New York Academy of Sciences.

Doel, Ronald, and Kirsten C. Harper. 2006. Prometheus unleashed: Science as a diplomatic weapon in the LBJ administration. *Osiris* 21: 66–85.

McNeill, John R., and Corinna R. Unger, eds. 2010. *Environmental histories of the Cold War.* New York: Cambridge University Press.

Milam, Erika, Deborah Coen, Amy Fisher, Cyrus Mody, Buhm Soon Park, and Jutta Schickore. 2020. Science diplomacy. *Historical Studies in the Natural Sciences* 50 (4): 323–481.

Miller, Clark A. 2006. "An effective instrument of peace." Scientific cooperation as an instrument of US foreign policy. *Osiris* 21 (1): 133–160.

Mooney, Jadwiga, and Fabio Lanza. 2013. *De-centering Cold War history: Local and global change.* London: Routledge.

Nye, Joseph S., Jr. 1990. Soft power. *Foreign Policy* 80 (Fall): 153–172.

Nye, Joseph S., Jr. 2004. *Soft power: The means to success in world politics.* New York: Public Affairs.

Phalkey, Jahnavi. 2013. *Atomic India: Big science in twentieth century India.* Ranikhet, India: Permanent Black.

Pietrobon, Allen. 2016. The role of Norman Cousins and Track II diplomacy in the breakthrough to the 1963 LTBT. *Journal of Cold War Studies* 18 (1): 60–79.

Rentetzi, Maria. 2018. A diplomatic turn in history of science. *Newsletter of the History of Science Society* 47 (1): 13–15.

Roberts, Geoffrey. 2020. Science, peace and internationalism: Frédéric Joliot-Curie, the World Federation of Scientific Workers and the origins of the Pugwash Movement. In *Science, (anti-) communism and diplomacy: The Pugwash Conferences on Science and World Affairs in the early Cold War,* eds. Alison Kraft, and Carola Sachse, 43–79. Leiden: Brill.

Rotblat, Joseph. 1967. *Pugwash—A history of the conferences on science and world affairs.* Prague: Czechoslovak Academy of Sciences.

Rotblat, Joseph. 1972. *Scientists and the quest for peace. A history of the Pugwash Conferences.* Cambridge: MIT Press.

Rotblat, Joseph. 1998. A social conscience for the nuclear age. In *Hiroshima's shadow,* eds. Kai Bird, and Lawrence Lifschultz. Stony Creek, CT: Pamphleteer Press.

Roxburgh, Ian W. 2007. Sir Hermann Bondi KCB. *Biographical Memoirs of Fellows of the Royal Society* 53: 45–61.

Rubinson, Paul. 2011. Crucified on a cross of atoms. Scientists, politics and the test ban treaty. *Diplomatic History* 35 (2): 283–319.

Rubinson, Paul. 2020. American scientists in "Communist Conclaves." Pugwash and anti-communism in the US. In *Science, (anti-)communism and diplomacy: The Pugwash Conferences on Science and World Affairs in the early Cold War,* eds. Alison Kraft, and Carola Sachse, 156–189. Leiden: Brill.

Sachse, Carola. 2020. Patronage impossible: Cyrus Eaton and his Pugwash scientists. In *Science, (anti-)communism and diplomacy: The Pugwash Conferences on Science and World Affairs in the early Cold War,* eds. Alison Kraft, and Carola Sachse, 80–117. Leiden: Brill.

Salomon, Jean-Jacques. 2001. Scientists and international relations: A European perspective. *Technology in Society* 23: 291–315.

Schwartz, Leonard E. 1967. Perspective on Pugwash. *International Affairs* 43 (3): 498–515.

Shields, Brit. 2016. Mathematics, peace and the Cold War: Scientific diplomacy and Richard Courant's scientific identity. *Historical Studies in the Natural Sciences* 46 (5): 556–591.

Skolnikoff, Eugene. 1993. *The elusive transformation. Science, technology and the evolution of International politics.* Princeton, NJ: Princeton University Press. A Council on Foreign Relations book.

Slaney, Patrick D. 2012. Eugene Rabinowitch, the Bulletin of the Atomic Scientists, and the nature of scientific internationalism in the early Cold War. *Historical Studies in the Natural Sciences* 42 (2): 114–142.

Smith, Tony. 2000. New bottles for new wine: A pericentric framework for the study of the Cold War. *Diplomatic History* 24 (4): 567–591.

The Royal Society of London. 2010. *New frontiers in science diplomacy. Navigating the changing balance of power.* London: The Royal Society of London/American Association for the Advancement of Science.

The Vienna Declaration. 1958. *Bulletin of the Atomic Scientists,* 14 (9): 341–344.

Turchetti, Simone. 2018. *Greening the alliance: The diplomacy of NATO's science and environmental initiatives.* Chicago: Chicago University Press.

Turchetti, Simone, Nestor Herran, and Soraya Boudia. 2012. Introduction. Have we ever been "transnational"? Towards a history of science across and beyond borders. *British Journal for the History of Science* 45 (3): 319–336.

Turchetti, Simone, Matthew Adamson, Giulia Rispoli, Doubravka Olšáková, and Sam Robinson. 2020. Just Needham to Nixon? On writing the history of "science diplomacy." *Historical Studies in the Natural Sciences* 50 (4): 323–339.

Turekian, Vaughan C. 2012. Science and diplomacy: The past as prologue. *Science and Diplomacy* 1 (1): 1–5.

Turekian, Vaughan C. 2018. The evolution of science diplomacy. *Global Policy* 9 (S3): 5–7.

Van Dongen, Jeroen, ed. (With Friso Hoeneveld, and Abel Streefland). 2016. *Cold War science and the transatlantic circulation of knowledge.* Leiden: Brill.

Voorhees, Jan. 2002. *Dialogue sustained. The multilevel peace process and the Dartmouth Conference.* Washington D.C.: US Institute of Peace Press.

Weisskopf, Victor F. 1969. The privilege of being a physicist. *Physics Today* 22 (8): 32–36.

Westad, Odd Arne. 2007. *The global Cold War. Third World interventions and the making of our times.* Cambridge, MA: Cambridge University Press.

Westad, Odd Arne. 2017. *The Cold War. A world history.* London: Penguin Books.

Whitesides, Greg. 2019. *Science and American foreign relations since World War II.* Cambridge, MA: Harvard University Press.

Wolfe, Audra J. 2018. *Freedom's laboratory. The Cold War struggle for the soul of science.* Baltimore: Johns Hopkins University Press.

Chapter 2
Moscow, Late 1960: A Breakthrough in East–West Dialogue

Abstract This chapter focuses on the sixth Pugwash conference, held in Moscow in late 1960. It argues, first, that this conference witnessed a decisive shift in terms of the quality of the dialogue about disarmament between US and Soviet scientists, giving rise to a novel form of technopolitical communication. Second, these exchanges carried new political weight because those around the Pugwash table included scientists close to the White House and the Kremlin. As a result, western perceptions of the Pugwash organization changed: US and UK governments came to see it as relevant to their interests. The Moscow conference also led to delegates from the US and the USSR creating a bilateral East–West Study group on disarmament, based on the Pugwash model but operating outside it. Overall, the chapter argues that this conference created an important platform for Pugwash and its scientists within the international realm of informal Track II diplomacy.

Keywords Techno-political communication · Arms control · General and complete disarmament · Soviet-American Disarmament Study Group (SADS) · Joseph Rotblat · Aleksandr Topchiev · Jerome Wiesner · Solly Zuckerman

2.1 The Path to Moscow: Maintaining Momentum, Raising the Profile of the PCSWA

After the exhilarating experience in Vienna, the fourth conference in late June-early July 1959 in Baden given to the theme of *Arms Control and World Security* was, in Rotblat's words, "not as effective" with the "process of reducing areas of divergence" both "slow and arduous" (Rotblat 1967, 23–24, 98–99). Although Rotblat viewed the fifth conference in August 1959, dedicated to *Biological and Chemical Warfare* as more successful, partly because there was broad agreement against the development and use of these weapons, he was nonetheless concerned about what he saw to be a loss of momentum since Vienna (Kaplan 1999; Perry Robinson 1998; Rotblat 1967, 24–26, 100–104). For their part Soviet scientists seemingly remained satisfied with how things were developing. This was, for example, apparent in the assessment of Pugwash by Alexander Vinogradov in 1959: "I do not know where else it is possible to express so freely one's views on the complex problems which arise as aspects of the atomic danger than at these international conferences" (Vinogradov 1959, 376–378). Vinogradov (1895–1975) was head of the Vernadsky Institute of Geochemistry and Analytical Chemistry, and for a time a regular at the PCSWA, attending the Baden, Moscow and (first) Stowe conferences in 1959, 1960 and 1961 respectively.

In an indication of the strengthening relationship between the scientists on the Continuing Committee, the decision was taken in autumn 1959 to hold the next conferences in superpower territory.[1] The sixth conference in 1960 would take place in Moscow—the first in a Communist country—whilst, to ensure "symmetry," the US would host the conference in 1961 (Rotblat 1967, 28–29).[2] This bold move was also perhaps strategic, providing a means to register the PCSWA more strongly on the international stage, but especially to raise its profile in Washington and London.

The Soviet Pugwash group swiftly set about organizing the Moscow conference which was, as would become the pattern in the Eastern bloc, sponsored by the country's Academy of Sciences (Evangelista 1999; Lüscher 2020, 2021). Initially arranged for April 1960, and then postponed until September, the sixth Conference finally took place between 27 November and 5 December.[3] The decision on a second postponement—taken in August, on the grounds of the upcoming US election—was very much against Rotblat's wishes. It was already a year since the fifth conference and he wanted to maintain momentum. As he explained to Eugene Rabinowitch, "if we put it off any further we shall have lost the impetus and any influence we may have had, not only with governments, but with scientists."[4] Rotblat was also keen to maintain good relations with Aleksandr Topchiev (1907–1962), who was emerging as a crucial colleague within the Continuing Committee. Topchiev, a petrochemist, was chief scientific secretary and vice president of the Soviet Academy of Sciences. For Eugene Rabinowitch he was, "the first Russian to bring (to Pugwash) his whole-hearted support."[5] Rotblat was worried that a further delay might generally "arouse much suspicion in the minds of our Soviet colleagues, and may make them feel that we have some ulterior motives." He was almost certainly also alarmed by the deteriorating relations between Washington and Moscow during 1960 in the wake of successive crises, most prominently the inaugural French nuclear test in the Sahara in February which aroused Soviet anger, amplified by the failure of the US to criticize the French at the UN, the U2 spy plane incident in May, and Khrushchev's breaking up of the Paris summit in June. All this took place against the backdrop of the faltering disarmament talks in Geneva. As Rotblat confided to Rabinowitch in August, the situation provided a real test of Pugwash:

---

[1] Minutes of the Continuing Committee, December 1959. RTBT 5/3/1/2 (Pt. 2).

[2] Minutes of the tenth meeting of the Continuing Committee, 21–23 June 1960. RTBT 5/4/2/17 (Pt 2). Harrison Brown to Harold Oram, 8 June 1961. RTBT 5/2/1/7 (11).

[3] Minutes of the meeting of the Continuing Committee, 21–23 June 1960, p. 2. RTBT 5/4/2/17 (Pt 2).

[4] Joseph Rotblat to Eugene Rabinowitch, 3 August 1960. RTBT 5/2/1/6 (32).

[5] Eugene Rabinowitch. 1970. The tasks of Pugwash. Opening speech at the twentieth PCSWA, Fontana, USA, September 1970. RTBT 5/2/1/20 (6).

If Pugwash is to be what it is supposed to be, then it is imperative that we should meet to discuss these problems when the political situation is so bad and there does not seem to be any ray of hope for some agreement being reached. Perhaps I am being too optimistic, but I had hoped that our Conference would be the means of gathering the threads again and provide the American and Soviet governments with a way for re-starting negotiations. This hope may not be fully justified, but I am firmly convinced that a postponement (…) will mean the end of the Pugwash Movement.[6]

In these private remarks, Rotblat reveals on the one hand, his ambitions for Pugwash to become part of the international disarmament conversation at the highest levels, and on the other, the fragility of an organization still finding its way in uncharted territory. Rotblat perhaps had another reason to avoid a further delay. It could, he thought, potentially jeopardize his bridge-building efforts with senior figures close to Whitehall, with which he personally had a strained relationship. Specifically, he had secured the participation in Moscow of Alistair Buchan (1918–1976), director of the newly established International Institute for Strategic Studies (ISS) in London, who was well-connected in senior political circles in the UK.[7] It remained the case that senior scientists and policy figures close to the British government tended to decline invitations to attend Pugwash conferences. For example, Guy Hartcup and Thomas Allibone report that British nuclear chief John Cockcroft "could not be persuaded" to go to Moscow because he thought the meeting would be of "little value" (Hartcup and Allibone 1984). Buchan could perhaps open the door to change: his colleagues at ISS included Anthony Buzzard, former head of Naval Intelligence, the Labour MP Denis Healey, the military historian Michael Howard, all of whom had Whitehall connections, and the former US Air Force Colonel Richard Leghorn (1919–2018) who was well-connected in Washington (Howard 2004).[8] Described as a "Chatham House for defence," the ISS enjoyed strong links to the political establishment and received financial support from the UK Foreign Office and the Ford Foundation. As Rotblat explained to Buchan, the further postponement until November reflected events across the Atlantic where several US colleagues who were involved in the Kennedy campaign feared that "participation in Moscow" could potentially cause them difficulties.[9] Such fears were a reminder of the cloud of suspicion hanging over Pugwash in the US. In the event, Buchan went to Moscow and would write a report about it that reached the British Foreign Office, where it would help to change opinions about Pugwash conferences being of "little value."

This inertia around the PCSWA within British scientific circles close to government was bound up with dislike of the organization in Whitehall. One problem was

---

[6] Joseph Rotblat to Eugene Rabinowitch, 3 August 1960. RTBT 5/2/1/6 (32).

[7] The son of Scottish-Canadian diplomat John Buchan (author of the novel *The Thirty Nine Steps*), and educated at Eton and Oxford, Alastair Buchan saw active service during World War II before working as a journalist for the *The Observer* prior to his appointment at the ISS in 1958.

[8] Buchan went on to an academic career in the field of International Relations, publishing widely on themes ranging from disarmament and various arms control treaties, Anglo-American relations, Henry Kissinger, and on Asia and China.

[9] Joseph Rotblat to Alistair Buchan, 29 August 1960. RTBT 5/2/1/6 (32).

Joseph Rotblat himself. In the mid-1950s, Rotblat—a physicist turned radiobiologist—had challenged the official position of the American and British governments that radioactive fallout from nuclear weapons tests posed little danger to human health (Higuchi 2020; Kraft 2018). Rotblat had changed direction in his career after the war, moving into the field of radiobiology: in his expert opinion, the evidence on these dangers remained unclear, but policy should err on the side of caution, and testing should be stopped. As he later reflected, in going public with his views he had "blotted his copybook" with the British government (Rotblat 1998, xxiii). But there were other problems. For example, a Foreign Office memorandum offering 'Guidance' on Pugwash just prior to the Moscow Conference stated that, "In our view, the Pugwash Conferences are potentially dangerous to Western interests, particularly in the disarmament negotiations, since the Communists may be expected to exploit these discussions for propaganda purposes," even if, as the report went on to concede, "presently the main inspiration remains technical, neutralist and non-political."[10] The Moscow conference had clearly caught the attention of Whitehall, triggering scrutiny of the PCSWA. The memorandum went on to emphasize the need to maintain a non-communist majority on the Continuing Committee and for "reliable and well-informed" scientists to dominate at its conferences so as to contain Soviet influence. How this might be achieved was not specified, but it indicates that Whitehall kept a watchful eye on the internal dynamics of Pugwash—both its British contingent, and on the international stage.

The point about the use of the PCSWA by the Soviets for propaganda purposes was not unfounded. Senior westerners within the organization had long been concerned about this—never more so than in the run up to and during the Moscow conference. As Rabinowitch acknowledged, Moscow would provide "an especially difficult test of the willingness (and capacity) of the Russian colleagues to avoid" using the PCSWA "propaganda purposes."[11] In the febrile political context of superpower tensions during 1960, the need to guard against anything that could in any way be construed as Soviet propaganda was seen as paramount. It had been agreed by the Continuing Committee that the Moscow conference would be private, but that a final statement would be issued publicly—contingent on sufficient unanimity and on what had been achieved. Concerns about the damage caused to Pugwash by adverse publicity contributed to a decision by the Committee to create the new post of Public Relations Officer to help foster a more favourable profile with the public, to which the British journalist and politically well-connected Wayland Young was appointed.[12]

---

[10] Foreign Office, *Guidance*, "Pugwash," No. 402, 23 November 1960. TNA F.O. 371/163160. Marked: confidential. The distribution list included UK High Commissioners in all Commonwealth countries and the British Ambassador in Dublin, Ireland. Note: The author is indebted to and wishes to thank Lawrence Wittner for depositing a copy of this primary source in the Joseph Rotblat Collection at the Churchill Archive Center, Cambridge, UK. RTBT 5/1/3/7.

[11] Eugene Rabinowitch. 1961. Thoughts on the meeting in Moscow. Draft Report, January 1961. RTBT 5/2/1/6 (39).

[12] Wayland Young (1923–2009), 2nd Baron Kennet, was a public figure in the UK: aristocrat, member of the House of Lords, one-time member of the Labour Party and later of the Social Democrat Party, and author of a book on the sexual revolution. The new public relations post was

2.2   Building Goodwill at the House of Friendship

When the conference finally began on 27 November 1960, John F. Kennedy had been elected the thirty-fifth president of the United States, bringing for many hopes of a new political era. The Moscow conference involved seventy-five delegates from fifteen countries, with the UK, US and USSR dominating numerically, collectively accounting for fifty-seven of those present. Other participants included four Chinese scientists—considered crucial amid tensions in Sino-Soviet relations—and two or three scientists from West and East European countries, including East Germany (Rotblat 1967, 105–108). Most striking, however, was the changed composition of the American delegation which, as Rabinowitch remarked, now included many "who had neither direct nor indirect association with previous Pugwash conferences," and who represented a "different part of the American scientific community."[13] He was alluding to colleagues with close connections to the White House typically forged through scientific advisory roles, most prominently the Harvard biochemist Paul Doty (1920–2011), the economic historian Walter Whitman Rostow (1916–2003), and the MIT engineer Jerome B. Wiesner (1915–1994).

Increased demand for scientific advice in government was a feature of the post-Second World War decades around the world. In the US, this demand expanded dramatically in late 1957 following the Sputnik crisis, most strikingly apparent in Eisenhower's creation of the President's Scientific Advisory Committee (PSAC) (Damms 2000, 57–58; Manzione 2000, 21–55; Macdonald 2015, 1–21; Rubinson 2011, 307–309, 2016). As Ronald Geiger has noted, the PSAC "placed scientists closer to the center of power than ever before," and was "at the height of its influence" during the Kennedy presidency (Geiger 1997, 354). In January 1961, Kennedy appointed Doty to the PSAC, made Wiesner his Science Advisor and appointed Rostow as deputy to his National Security Advisor, McGeorge Bundy. This trio came to count amongst Kennedy's closest advisors in a White House that was receptive to and placed value on the views of scientists (Wiesner 1963). Also present in Moscow were other members of what Bernd Kubbig has called the "East Coast arms control establishment" then "coming into being with its centers at MIT and Harvard," including Donald Brennan, Bernard T. Feld, David Frisch, Thomas Schelling and Louis B. Sohn (Kubbig 1996, 6; Rubinson 2020). The participation in Moscow of this "different part" of the scientific community lent new political weight and meaning to the conference and, by inference, to the wider Pugwash organization.

As scientific advisors to the Kennedy White House, Doty, Rostow and Wiesner acquired a degree of public prominence. Indeed, in 1967, the trio featured in an article in *Time Magazine* by Theodore H. White highlighting the extensive involvement of

---

funded by Lord Simon of Wythenshawe. RTBT 5/4/2/14. In 1962, after a serious falling out with Rotblat, Young was replaced by Gerald Leach and, in turn, the Welsh journalist John Maddox, then science writer at *The Manchester Guardian*. Maddox later became a long-serving editor of the journal *Nature*. RTBT 5/4/6/6 (Pt 1).

[13] Eugene Rabinowitch. 1961. Thoughts on the meeting in Moscow. Draft Report, January 1961. RTBT 5/2/1/6 (39), 1.

Harvard and MIT scholars in national affairs (White 1967). For White, these men were the "Action Intellectuals," a "brotherhood of scholars" who had "left their quiet and secure niches on the University campuses" and now represented a "new power system in American life." In 1960, these men came to the Pugwash conference in Moscow because it provided a rare—and for many of them first—opportunity for face-to-face conversations with senior Soviet scientists. The PCSWA had succeeded in drawing into its orbit scientists close to the US government.

Rabinowitch was encouraged equally by the "outstanding" Soviet delegation in Moscow. The make-up of the delegations of the superpowers for each Pugwash conference was the result of a painstaking process managed within the Continuing Committee involving careful calibrations about matching delegations in terms of seniority, both scientifically and politically. In fact, Soviet scientists close to the Kremlin had always attended Pugwash conferences. Two of the most prominent were Aleksandr Topchiev and the geophysicist Evgenii K. Federov (1910–1981), Secretary of the Soviet Academy of Sciences and a member of the Soviet test ban negotiating team (Lüscher 2020, 128). Both had been members of the Continuing Committee since 1958. In 1960, the Soviet list included newcomers also close to the Kremlin. These included Vladimir Arzumanyan, head of the Institute for World Economic and Foreign Affairs and a disarmament expert; elder statesman of science Anatoli A. Blagonravov (1894–1975); and the metallurgist Vasily Emelyanov (1901–1988), head of the Institute for the Utilization of Atomic Energy and Soviet representative to the International Atomic Energy Agency. As hosts, the Soviets were perhaps especially keen to field a strong array of politically well-connected scientists: also present were the historian Vladimir Khvostov (1905–1972), the physicist and Nobel laureate Igor Tamm (1895–1971), and the military general and editor of the *Red Army Magazine* Major General Nikolai A. Talensky (1901–1967). Talensky was an influential link between the Ministry of Defence, civilian institutes and the Disarmament Committee of the Soviet Academy of Sciences and he would later become a key player in formulating the Soviet strategy on anti-ballistic missiles (Kubbig 1996).

The Moscow conference therefore brought together scientists involved in the disarmament conversation at the highest level in both Moscow and Washington. Meanwhile, the British delegation included the usual stalwarts, Rotblat, Cecil F. Powell (1903–1969), Rudolf Peierls (1907–1995), and the British-Canadian Quaker and Labour MP Philip Noel-Baker (1889–1982) who had just been awarded the Nobel Peace Prize for his "longstanding contribution to the cause of disarmament and peace" (Laucht 2012; The Nobel Peace Prize 1959). In addition to Alistair Buchan were two other British newcomers to Pugwash, the cancer clinician Alexander Haddow (1907–1976), and former President of CERN Ben Lockspeiser (1891–1990), both of whom, through their various government advisory roles, were politically well-connected in Whitehall (Bergel 1977; Edwards 1994).

The conference took place in the *House of Friendship,* a former Tsarist palace now used as a venue for international events.[14] The format resembled that of an academic scientific conference with the program centered around papers read in

---

[14] Documents relating to the Moscow Conference. RTBT 5/3/1/3.

plenary sessions, with ensuing discussions taking place around a large U-shaped table with simultaneous translation in Russian and English, in this period the two languages of the PCSWA.[15] The Russians provided a small team of translators—translation being in itself a form of transnational practice, with trust in the reliability of translations being essential to the overall building of confidence in the integrity of international exchanges. That said, some of the key western participants could operate in English and Russian, notably Eugene Rabinowitch, who was born and grew up in St. Petersburg—a powerful asset for the PCSWA in terms of relations with Soviet colleagues.

As had become customary, the conference opened with messages from dignatories, including Khrushchev and Bertrand Russell (1872–1970). Papers by Aleksandr Topchiev and Paul Doty followed (Doty 1960; Topchiev 1960). In his paper, entitled 'Disarmament: the most urgent problem of today,' Topchiev reminded the audience of important world events since the Baden conference in June 1959, warning that "nuclear war may break out at any moment" (Topchiev 1960, 21–32). In a thinly veiled reference to the U2 episode and swerving briefly into propagandistic rhetoric, he went on, "We had occasion to see how rudely the time-honored sovereign rights of states, regarded as norms of international law, were trampled upon, how unlawfully their air space was violated by foreign airplanes. The airplanes which sneaked over alien skies carried equipment for military espionage. They can also carry nuclear bombs. We hear words that sow poisonous seeds of mistrust in the relations between countries and containing appeals for arms drive and aggression" (Topchiev 1960, 24). Perhaps this was not unexpected; if Topchiev was collegial within the Continuing Committee, he kept to the Moscow line in his papers at conferences. As Fabian Lüscher has noted, the Soviet scientists remained fully aware that "Speaking Bolshevism" was essential for their relations with their various domestic audiences and key to the continuation of the Pugwash project in the Soviet Union (Lüscher 2020, 2021). Perhaps Topchiev and his colleagues were keen to hold on to whatever degree of autonomy and agency this afforded them at the Pugwash table. Westerners were no doubt aware of this too, and understanding this was seen as part and parcel of the 'rules of engagement' that were being established in Moscow. At any rate, Doty ignored the side-swipes, sticking to his paper that presented a detailed survey of the problems of disarmament—which the Soviets "disliked," presumably because it emphasized the Arms Control approach developed by Doty and his East Coast colleagues, and which prompted "heated discussions" (Doty 1960; Kubbig 1996, 16). Perhaps as early emissaries for the incoming Kennedy administration, the Americans were prepared for and ready to overlook instances of saber-rattling, staying focused instead on their own goals, not least learning more about the Soviet mindset, and Soviet perceptions of the disarmament process.

1960 had been a busy and turbulent year in Geneva where, between March and June, both superpowers had tabled disarmament proposals in quick succession which had been discussed and rejected (Bulletin of the Atomic Scientists 1960, 336–339).

---

[15] Documents relating to the Moscow Conference. RTBT 5/2/1/6 (42).

The faltering negotiations—beyond the Cold War grandstanding and obstinacy—reflected a fundamental lack of understanding conceptually and culturally between the two sides, and their very different framing of and approaches to disarmament. The Moscow conference program involved panels of papers given to in-depth analysis of the myriad problems of disarmament from both the American and Soviet viewpoint. In the opening session, Noel-Baker and Khvostov surveyed the history of the Geneva negotiations, whilst the US lawyer Louis B. Sohn (1914–2006), gave two papers, a historical survey of arms control, and a joint paper with David H. Frisch entitled 'Arms Control in the 1960s' setting out the idea of disarmament "zones" (Sohn and Frisch 1960; Pasqualucci 1998, 924–944).[16] This panel prompted lengthy discussion—setting the pattern for much of the conference. Across the nine days, most of the Americans gave papers in Moscow, including Donald Brennan, Bernard Feld, and Jerome Wiesner, as did the Soviets, including Federov, Talensky and Blagonravov (Wiesner 1960; Brennan 1960; Feld 1960; Federov 1960; Talensky 1960; Blagonravov 1960). In-depth discussions ranged across both the technical challenges and the political stumbling blocks bedeviling the official negotiations in Geneva; sometimes there was agreement, for example, shared concerns about the dangers of accidental war and about the spread of nuclear weapons. There were differences on chemical and biological weapons. This issue was high on the Soviet agenda, with several papers calling for a ban on their development and use (Dubinin 1960; Imshenetski 1960).

Moscow provided an early international platform for advancing the US 'Arms Control' paradigm, an approach to disarmament pioneered and strongly favored by those Americans seated around the Pugwash table. Indeed, the Americans felt generally that they were more active than the Soviets when it came to research into and the theorization of disarmament.[17] The Arms Control approach emphasized the "means" of disarmament rather than the "ends," essentially conceiving nuclear disarmament as a process which involved, for example, verification and inspection procedures. Wayland Young also picked up on this, noting in his report on the conference his impression that the Americans thought the Russians "had not done their homework as well as it is done at Harvard."[18] For example, earlier in the year, the nascent American Arms Control community had held a summer school, the work and findings of which were summarized in Moscow by Feld (1960, 461–469). This had been organized by the American Academy of Arts and Sciences (AAAS) and the Federation of American Scientists—specifically the Committee on the Technical Problems of Arms Limitation—and financed by the Twentieth Century Fund. Papers from this meeting—by Brennan, Doty, Feld, Kissinger, Sohn, Wiesner and others—were

---

[16] Born in Lemberg/Lviv in 1914, Sohn became known for his theorization of disarmament and disarmament negotiations and in the development of international law relating to this topic; later he also gained a reputation as a respected and influential advocate for and practitioner in international Human Rights.

[17] Transcript of roundtable discussion, 'Report from Moscow on disarmament and world security,' WGBH-TV, 3 January 1961. Transcript reproduced in Appendix V, the Dodd Report, 107–125.

[18] Wayland Young. The sixth international PCSWA, Moscow, 27 November–4 December, 1960. RTBT 5/2/1/6 (38). The official dates of the conference were 27 November until 5 December. The reason for the discrepancy in the end date of Young's report is not known.

published in the AAAS journal *Daedalus* in September 1960 (Daedalus 1960; Edsall 1960, 791; Garwin and Geiduschek 2013).[19] The papers in this volume of *Daedalus* did much to establish and define the new concept of 'Arms Control' and carried a great deal of weight in the Kennedy White House (Rubinson 2011). For the Soviets, the Moscow conference afforded an introduction to the Arms Control paradigm and to those who had conceived it. Briefly stated, this paradigm was premised on a schedule of arms reduction requiring verification and inspection, which was incompatible with the Soviet 'general and complete' approach to disarmament. At this juncture, Arms Control was anathema to the Soviets and rejected by the Kremlin.

The papers on the Moscow program constituted at the time a groundbreaking body of work in the analysis and theorization of disarmament. This exchange of ideas and views exposed and helped to clarify the differences between the positions of the superpowers. Grasping these differences was a step towards better understanding between the Cold War adversaries. As Jerome Wiesner remarked after his paper:

> One of the most important aspects of Pugwash Conferences was the possibility of meeting, arguing, disagreeing, and yet feeling that differences came from differences in points of view, and continued discussions could reveal the reason for disagreement and what steps could be taken to get closer together. (Wiesner 1960, 210)

Here, Wiesner captured precisely the process that the Pugwash project sought to promote. Whatever the views in the Kremlin and in the White House about the PCSWA and about what scientists at Pugwash conferences could do within the realm of disarmament, those present in Moscow took their work very seriously. As it turned out, the incoming Kennedy administration—with limited foreign policy experience and keen to gauge the climate in the Kremlin—was very interested in hearing news of the Moscow conference. Immediately on their return to the US, Rostow and Wiesner reported on the conference to the White House, the State Department and the Pentagon, with Kubbig noting that "Wiesner at least had been able to deliver policy-relevant information from the US administration to the Soviets; in Moscow, US Ambassador Thompson had also asked other American conference participants to do the same" (Kubbig 1996, 9). This was exactly a chain of events envisioned by the founders of the PCSWA: namely, the content of East–West discussions about disarmament made possible by Pugwash being relayed to senior political and policy-making circles of the nuclear superpowers.

Progress in Moscow was aided by a decision announced by Rotblat during the fifth session on 30 November regarding an impromptu change to the program (Rotblat 1960, 292; 1967, 28). On two afternoons the delegates would break out into smaller groups specifically with a view to facilitating more detailed "informal talks" to be held in the congenial location of the Metropole Hotel. This innovation proved highly successful and seeded the idea for 'Working Groups' which, beginning in 1961 in Stowe, became an integral part of the conference program. The importance of these groups, which typically involved between ten and fifteen scientists, as sites of private discussion and building understanding between scientists cannot be overstated. For

---

[19] The AAAS comprised three committees: Studies in Arms Limitation, International Conversation, and Public Responsibility of Scientists—chaired respectively by Bernard Feld, Leo Szilard and Eugene Rabinowitch.

example, as early as 1962, Eugene Rabinowitch observed that the success of the Conferences was measured not by consensus, which was not always reached, "but by the feeling of the participants that, in its sessions and even more in its working groups and private conversations, certain prejudices had been destroyed and certain ideas put across."[20]

Moreover, in the course of the exchanges in Moscow—which were variously probing, antagonistic, sceptical, conciliatory but always cautious—human relationships were developing. This human dimension was further aided by a busy social program, including nights at the Bolshoi, and dinners hosted by Topchiev at his home: here too were opportunities for private discussions out of the spotlight. In this way, a qualitative change in the cross-bloc conversations took place, and a path towards developing mutual understanding between those present opened up and a degree of goodwill was established. From this came the possibility of overcoming mistrust and replacing it with trust—albeit tentative and contingent.

2.3   Three Western Perspectives on the Moscow Conference

Three westerners—Alistair Buchan, Eugene Rabinowitch, and Wayland Young—produced reports on Moscow which afford valuable insights into what took place.[21] Unsurprisingly scientists from both superpowers adhered to the respective 'red' lines at the official disarmament talks on-going in Geneva. The divergent approaches of the US and USSR to the disarmament problem, including the gulf between general and complete disarmament (GCD: the Soviet position) and reduction/controls (the American position), was everywhere apparent. But sensitive issues were tackled, for example Rostow would later recall that almost fifty per cent of time was spent on the intractable problem of inspection—which remained a major stumbling block in Geneva.[22] For all these fundamental differences, Wayland Young was struck by the willingness to listen to and consider the viewpoint of others. There were jokes too, for example, Young reported a discussion of how "arms control is almost a dirty word to the Russians, and "stable deterrent" is a full-blown dirty word. There was laughter about the corresponding possibility that "general and complete disarmament" might become a dirty word in America."[23] Moments of light-hearted banter point to the

---

[20] Eugene Rabinowitch. 1962. The future of COSWA. RTBT 5/2/1/11 (21).

[21] Eugene Rabinowitch. 1961. Thoughts on the meeting in Moscow. Draft Report, January 1961. RTBT 5/2/1/6 (39). Buchan, Alistair. *Notes on Pugwash*, Moscow, 26.11.1960–5.12.1960. RTBT 5/2/1/6 (39). Wayland Young, The sixth international PCSWA, Moscow, Nov. 27th–Dec. 4th, 1960. RTBT 5/2/1/6 (38). Marked confidential, Young's report included the caveat: "The following notes have been prepared by Wayland Young, and are confidential. They are in no sense an agreed account of what happened," making clear that this was not an official Pugwash document.

[22] Comment by Walter Rostow. Round Table discussion about the Moscow Conference broadcast by WGBH-TV, Boston, 3 January 1961, 118. Transcript included as Appendix in the Dodd Report, 1961, 107–125.

[23] Wayland Young. The sixth international PCSWA, Moscow, Nov. 27th–Dec. 4th, 1960. RTBT 5/2/1/6 (38), 7–8.

collegial and congenial atmosphere created between the participants, but there were also moments of bristling tension. As Young also wrily noted, the first day was "devoted to courtesies, the second and third to distrust and disharmony." Tensions were everywhere present to serve as a reminder, if any were needed, that this was a Cold War powerplay, with those present engaged in delicate frontline diplomacy by 'back-channel' means, knowing that what was said would find its way to their respective governments. Indeed, this was the hope and the point.

In his report, Young identified Doty, Richard Leghorn, Rostow and Wiesner as the four "most influential" Americans. In his view, the Soviets saw Jerome Wiesner as the key American player, as he put it, "definitely the man they thought they were talking to, and, as far as getting things done goes, I should guess they were right."[24] That is to say, the Soviets perceived Wiesner to have the ear of the incumbent President and therefore the chief channel to the White House (Rosenblith 2003). As for the Soviets, Young portrayed Topchiev as "the general smoother-over, invitation-conveyor and ring-holder," whilst "Kapitza's knowledge of the western mind was again and again invaluable, especially at tense moments."[25] But in his view, it was Evgenii K. Federov who "made the running."[26] Tellingly, Young described the development of a Federov-Wiesner axis, with the two men "sat opposite each other for the first five days," and that "a great deal of communication was going on between them all the time, a real conversation and meeting of minds."[27] This description of transnational cross-bloc exchange in action points to the new and distinctive kind of informal diplomacy made possible by the Pugwash project.

Young also remarked on what he viewed to be a sudden and decisive change on the fourth day (Thursday 1st December) when the Nobel Prize winning chemist Petr Kapitza (1894–1984) used a diagram to discuss a disarmament problem and "thenceforth the blackboard was in vogue and the barriers were down." Subsequently, in Young's view, the real "catalyst of the whole conference was another diagram put up by Jerome Wiesner"—the "Wiesner Diagram"—which was "redrawn every day from then on."[28] For Young this was a "breakthrough moment" which immediately engendered a more relaxed atmosphere. The Wiesner Diagram (Fig. 2.1) plotted disarmament against an axis A, for the level of arms of any given type, and an axis T, for time. "The argument between the two sides is thus about the angle of incidence of the rising line I for inspection. The Russians want it flat, the Americans steep."[29] The difference between the two lines represented the space in which verification and inspection would take place—where there was major disagreement. That is to say, the graph visualized the incompatible approaches of Arms Control (the US approach) and disarmament (the 'general and complete' model of the Soviets). Young

---

[24] Ibid, 2.

[25] Ibid, 3.

[26] Ibid, 1.

[27] Ibid, 2.

[28] Ibid, 3–4. The Wiesner diagram is also reproduced in the *Proceedings of the 6th Pugwash Conference*. Moscow November 1960, 298.

[29] Ibid, 8.

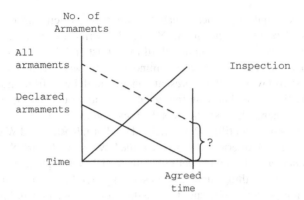

Fig. 2.1 Wiesner Diagram, Moscow, 1960

The graph charts armaments as a function of time. When the descending curve of declared armaments in a given category reached zero at the agreed time, there must be mutual certainty that the descending curve of all armaments also reached zero. If it did not, the ratio of all armaments to declared armaments would be infinite, and this would be dangerous. The maintenance for a time of an agreed low level of armaments on both sides would ensure that this ratio would be finite; perhaps 2, or 5. This would be less dangerous. At some time inspection must reveal this ratio.

Source: *Proceedings*, p. 298. PCSWA 6, Moscow. 30 November, morning, session 7.

reported that the validity of this diagram was "fully accepted by both sides."[30] The graphical representation of a technical—but also political—problem provided an innovative framework for analyzing the problem of inspection, one that intellectually created a new 'space' to probe and identify room for maneuver and possibilities for compromise. Diagrams and graphs, concepts and correlations: here was a language that scientists shared and understood. Here was the basis for a different style of technopolitical communication and exchange. For Young, this mode of working paved the way for all "the real progress" that was made during sessions focused on the problems of inspection and control—where technical expertise could play a major role.[31]

In his report, Eugene Rabinowitch viewed Moscow as "overall a success." He was especially relieved that, for the most part, the Soviets had toned down their rhetoric and "scrupulously adhered to their promises of privacy, restraint on press reporting, and avoidance of identification of the conference with official Soviet policies for propaganda purposes."[32] Rabinowitch also highlighted an important double move in Moscow—improved mutual understanding and enhanced access to government—which, in his view, had made this conference so important for the future of Pugwash. As he put it:

---

[30] Ibid, 9.

[31] Ibid, 8. Papers given on the fourth day included, for example, those by Bernard T. Feld, N. A. Talensky, David Inglis, and the Cornell physicist Jay Orear.

[32] Eugene Rabinowitch. 1961. Thoughts on the meeting in Moscow. Draft Report, January 1961. RTBT 5/2/1/6 (39), 4–5.

> While our capacity to bring about a frank exchange of opinions in the disarmament field among competent scientists who have the trust of their governments is, as of today, our most important achievement, it is not our only achievement. We have established a wide—even if vague—mutual understanding and community of attitude, and thus we are able to provide a wide-open unofficial channel for indirect exchange of opinions between governments.[33]

The "community of attitude" referenced the growing stock of goodwill and a deepening of relations between Pugwash scientists. This was a vital resource for an emerging *modus vivendi* between American and Soviet scientists which, crucially, provided a framework within which "unofficial"—informal—diplomacy became possible. In Moscow the means by which the organization could best pursue and realize its aims was taking shape. Others present in Moscow likewise point to its significance including, for example, in the report filed by the West German Eckart Heimendahl, the later recollection of the East German physicist Heinz Barwich that this conference was "the big one," and Walter Rostow's later assessment that this conference had been especially important for building contacts with the Soviets.[34]

The third report on Moscow was that by Alistair Buchan, which, in addition to providing another perspective—and that of a Pugwash outsider—proved significant in the UK where it reached Whitehall and alerted senior political circles there to the changing nature of the conversations at the PCSWA. Moscow was Buchan's first PCSWA, and the paper he gave addressed the theme of "Surprise attack" (Buchan 1960). As he conceded, he had neither been to Russia or "talked at length" with Russians before. His analysis mixed scepticism with a sense of being positively surprised. Noting its "rather different character," he concluded of Moscow that "this may have been not only the most important Pugwash meeting so far, but the most political in content."[35] Describing a Pugwash "spirit or tradition" that meant that "the meetings take the form of discussion rather than dialectic," Buchan quickly discerned the 'rules of engagement' operating around the table, noting the seemingly unspoken agreement that "neither side should attempt to press an argument or advantage too far, since this line of private communication must at all costs be kept open" (Kraft and Sachse 2020).[36] Buchan wondered how close the Russians at the *House of Friendship*

---

[33] Eugene Rabinowitch. 1961. Thoughts on the meeting in Moscow. Draft Report, January 1961. RTBT 5/2/1/6 (39), 4.

[34] For the recollections of Heinz Barwich, see his testimony before the US Senate in 1964, three months after his dramatic defection to the US. *Nuclear scientist defects to the US.* Hearing before the Subcommittee to Investigate the Administration of the Internal Security Act and other Internal Security Laws of the Committee on the Judiciary. US Senate. 1965. 89th Congress, First Session, 15 December 1964. US Government Printing Office, Washington. Heimendahl, Eckart. 1960. 'Report on the sixth PCSWA, Moscow, 1960.' *VDW Rundbrief* 8. File B43.12, Auswärtiges Amt, Berlin. Interview with Walter Whitman Rostow by Richard Neustadt, 11 April 1964. John F. Kennedy Library Oral History Program: 28–30.

[35] Alistair Buchan. 1960a. Notes on Pugwash, Moscow, 26.11.1960–5.12.1960. RTBT 5/2/1/6 (39), 1. Buchan's comments raise questions about the basis for his assessment, given that until now he had not been to a conference. One possible explanation is that he moved in political and policy circles in Whitehall where he was likely privy to conversations about the PCSWA.

[36] Alistair Buchan. 1960a. Notes on Pugwash, Moscow, 26.11.1960–5.12.1960. RTBT 5/2/1/6 (39), 1.

were to the Kremlin, was puzzled by what he called the sudden Soviet "passionate desire for speedy disarmament," and described terse exchanges involving Talensky and Federov on the topic of inspection—which, he noted, had been smoothed by Wiesner.[37] He also detected deep Soviet unease about China and West Germany. Tellingly, Buchan noted that the "most valuable part" of the conference was "what was said in private"—although he did not elaborate further on this. Overall, however, in his view, the two sides were and remained "poles apart" from the start with little "headway" made.[38]

Buchan's perceptive but downbeat interpretation of Moscow contrasted with the more optimistic and positive analyses of Young and in particular Rabinowitch who, as a veteran Pugwash insider, had more experience of working with the Soviets which perhaps engendered different and/or more modest expectations of the conference. Taken together, the three accounts paint a vivid picture of the probing encounter between East and West around the conference table, of the close engagement with the American and Soviet positions on the disarmament process, of the cut and thrust of the discussions, and of the different personalities involved. The depth and range of engagement with disarmament problems is strikingly apparent in the *Proceedings* of the Moscow conference which stand as a remarkable treatise on the conceptions and theorization of disarmament on both sides of the bloc divide at this point in the Cold War.

2.4 Reflections of the East Coast Elite

Improved understanding, changing attitudes, the nature of relationships and the existence of both trust and mistrust, are by their nature elusive and therefore difficult to assess, measure and substantiate. But some insights into these 'human' aspects as experienced by some of the Americans in Moscow can be gleaned from a televised roundtable discussion of the conference broadcast in early January 1961 featuring the East coast quintet of Paul Doty, Richard Leghorn, Alexander Rich (1924–2015), Walter Rostow, and Jerome Wiesner and chaired by MIT Professor of Economics, Lawrence Martin.[39] Indeed, Wiesner joked about Boston being a "hotbed of peace-mongering."[40] In clarifying why they had travelled to Moscow, Rostow explained that "we were seeking really to define rather precisely what the points of disagreement were" and emphasized that the conference had provided a valuable opportunity to learn about and better understand the Russians. Doty made clear that "we were there

---

[37] Soviet speediness was explained by Buchan in terms mainly relating to their leadership in rocketry and the increasing effects of the arms race on the Soviet economy.

[38] Alistair Buchan. 1960a. Notes on Pugwash, Moscow, 26.11.1960–5.12.1960. RTBT 5/2/1/6 (39), 3.

[39] Transcript of roundtable discussion, 'Report from Moscow on disarmament and world security,' WGBH-TV, 3 January 1961. Appendix V, the Dodd Report, 107–125.

[40] Transcript of roundtable discussion, 'Report from Moscow on disarmament and world security,' WGBH-TV, 3 January 1961. Appendix V, the Dodd Report, 107.

as individuals, not representatives of the government, although I think we carried with us the confidence of the government," adding that "many of us have been busy since returning, communicating and transmitting information to our own government. The impact of the conference, I think, has been somewhat more than we expected."[41] Noting the "unfortunately somewhat propagandistic" nature of Topchiev's paper, Doty estimated that there had been a daily contact time of ten-twelve hours for the duration of the conference, recalling too evenings talking over dinner at the homes of their hosts, including Topchiev, whilst Wiesner and Rich talked about visits they had made to Soviet laboratories.[42] The panel went into considerable technical detail about the different American and Soviet positions on disarmament, pithily summarized by Wiesner as: "We would like a lot of inspection before we do very much disarming, and they would like a devil of a lot of disarmament before they permit much inspection." The panelists made clear that there had been forthright exchanges with the Soviets, as Rostow put it, "There was no American anxiety, fear, belief, accusation that wasn't quite temperately but firmly aired there."[43] The panelists felt they had come away from Moscow with new insights into the Soviet mindset which, in turn, had accorded them better understanding of their conception of and approach to the disarmament process. In the closing part of the roundtable, Martin raised the issue as to whether the Soviet scientists had influence in the Kremlin. Leghorn said he was not sure "how much they represented governmental thinking," whilst Rostow thought the Kremlin might consult with the scientists they had met in Moscow, but that they were not the makers of Soviet disarmament policy. He went on, with an eye to the imminent change in the White House, "I think it's a high duty of the next administration to find out what the specific gravity of this element is."[44]

The transcript of this roundtable was included as an appendix to the Dodd Report when this was published in 1961—along with the text of Aleksandr Topchiev's paper in Vienna in 1958, and a copy of an article reporting on the Moscow conference by Louis B. Sohn published in *The Nation* on 14 January 1961.[45] In this article, Sohn reflected that he had learned a great deal about Soviet thinking and about their calculations in approaching nuclear disarmament, adding that he thought them more flexible than was generally understood. This was anathema to Dodd, who responded that Sohn was overly sympathetic to the Soviets and, as such, naive; he was dismissive of the views expressed in the TV roundtable and scathing about the "unrealistic idealism" of the PCSWA and its scientists generally.

---

[41] Transcript of roundtable discussion, 'Report from Moscow on disarmament and world security,' WGBH-TV, 3 January 1961. Appendix V, the Dodd Report, 107–108.

[42] Ibid, 110.

[43] Ibid, 110, 124.

[44] Ibid, 124–125.

[45] Louis B. Sohn, The Nation, 14 January 1961. Reprinted in Appendix IV, The Dodd Report, 104–106.

2.5   A Tangible Outcome: An East–West Study Group on Disarmament

At the twelfth meeting of the Continuing Committee in London in early March 1961, there was agreement that Moscow "had had the greatest impact on public opinion and in official circles so far achieved by a Pugwash conference."[46] Viewed in hindsight within Pugwash, the Moscow conference took Pugwash engagement with disarmament to a new level.[47] Certainly, this experience boosted confidence within the leadership about what the PCSWA might contribute to the disarmament process and provided a platform on which it hoped to build. There had been a qualitative change in terms of the depth of engagement and reciprocal exchange across the range of disarmament problems and about the divergent approaches to disarmament. Moscow was important in terms both of establishing the 'rules of engagement' between American and Soviet scientists for tackling disarmament and marked a step forward in the trust-building process between them.

That said, establishing trust remained the most difficult challenge facing the organization. Of course, all were acutely aware that in Moscow the existence of mistrust was, as Eugene Rabinowitch put it, "incontrovertible," emphasizing too what everyone knew, that without trust disarmament was not possible (Rabinowitch 1960, 689). On the Soviet side, Aleksandr Topchiev similarly declared that Pugwash needed first "to consider ways to increase trust" and, to this end, the conference "should stop worrying about cheating and deception and worry about understanding" (Topchiev 1960, 396–397). As Topchiev remarked, those present trusted each other as scientists and as researchers: in this, he saw a starting point. The challenge before them lay in finding ways to emulate within Pugwash the trust operating between scientists in the academic setting.

But of course, the disarmament process and the Cold War intersection between science and politics constituted a very different context. All recognized that they were moving into uncharted terrain; all were aware that their governments were watching. Scientists might share a common language, methods of working, and an intellectual outlook/mindset—an oft-repeated Pugwash narrative—but this alone was insufficient for the Pugwash project to succeed. Understanding and a degree of trust were needed. One indication that steps in this direction had been taken in Moscow was an agreement between the Soviets and Americans to take forward an idea mooted by the American mathematician Donald Brennan, then at the MIT's Lincoln Laboratory, about setting up an East–West study group on disarmament

---

[46] Minutes of the twelfth meeting of the Continuing Committee, London, March 1961. RTBT 5/3/1/2 (Pt 1) (3). At the fourth meeting of the EPG in April 1961 when Joseph Rotblat updated his European colleagues on this Continuing Committee meeting, he reported too that the Soviets had behaved themselves at the Moscow Conference. RTBT 5/2/3/4.

[47] For the program, see the *Proceedings of the 6th Pugwash Conference in Moscow,* November 1960.

(Kubbig 1996, 12–14).[48] The first steps were taken in bilateral discussions immediately after the conference and, shortly thereafter, a special committee of the AAAS, chaired by Brennan, prepared a detailed proposal for the project.[49] At this point, the Soviet Academy of Sciences and the AAAS agreed to co-sponsor the new study group. Initial plans for European participation gave way to a bilateral framework: after lengthy discussions, the East–West group—known also as Soviet American Disarmament Study Group (SADS)—met for the first time in June 1964 (Rabinowitch 1965, 13; Kubbig 1996).[50] Between 1960 and 1964, this group—which came also to be known colloquially in the US as the 'Doty group'—took shape in lengthy negotiations, with Soviet physicist Mikhail D. Millionshchikov (1913–1973) and the MIT biochemist Paul Doty as co-chairs, and with funding from the Ford Foundation (Kubbig 1996, 15–17).[51] The inaugural meeting of the SADS group took place in June 1964 at Harvard University—with the organization continuing until 1975.[52]

2.6 Not Playing the Game: The Chinese in Moscow

If there were positive developments in US-Soviet relations at the Moscow conference, by contrast, there was alarm amongst western members of the Pugwash leadership about the behavior and attitude of the Chinese scientists (Barrett 2020). Westerners looking for signs of Sino-Soviet tension were disappointed. On the contrary, in his paper the leader of the Chinese delegation, physicist Zhou Pei Yuan, reaffirmed the friendship between China and the Soviet Union. Zhou Pei-Yuan reserved his anger for the US for, variously, pursuing a "policy of war and aggression," for obstructing the disarmament process, for disparaging the government and people of the Soviet Union, and its peace policy, and for trying to "find a rift between the two great socialist

---

[48] Kubbig reports that Brennan's suggestion chimed with American frustrations about the rising numbers and influence of non-natural scientists in the Soviet delegations at the PCSWA and which they linked with a "strengthening ideological-propagandistic faction" that was working against "serious discussions between technicians."

[49] The idea for the disarmament study group was "warmly welcomed and endorsed" by the Continuing Committee. Minutes of thirteenth meeting, 2–15 September 1961, Stowe, VT. RTBT 5/3/1/2 (Pt 1) (3).

[50] Notes on the fifth meeting of the European Pugwash Group, 6–8 October 1961, Geneva. RTBT 5/2/3/5. The French Pugwash Group, especially Jules Moch was, for example, also spurred by Moscow to take up its own studies on disarmament. B. T. Feld to R. Leghorn, 30 April 1961. RTBT 5/2/3/4.

[51] Report on second plenary session, 13 September 1964. East–West Study Group, chair: Bentley Glass. Reports by Doty and Millionshchikov. Proceedings of the 13th Pugwash Conference in Karlovy Vary, 1964, 71–73. Shepard Stone was involved in the process of securing funding from the Ford Foundation—which came with the stipulation that Paul Doty was given the role of deciding on membership of the group. Donald Brennan provided regular updates to the Pugwash Continuing Committee about the SADS negotiations.

[52] The first SADS meeting was reported upon briefly at Continuing Committee meeting no. 19, Karlovy Vary, September 1964. RTBT 5/3/1/2 (Pt 1) (4).

countries China and the USSR which are bound by great friendship" and/or "drive a wedge" between them (Pei-Yuan 1960, 517–521). Moreover, he expressed great disappointment with the American delegates in Moscow for holding views "detrimental to the cause of world peace" whilst vehemently rebuking Jerome Wiesner for his inability to call the People's Republic of China by its full and proper name, and for trying to stir trouble over Taiwan (Pei-Yuan 1960, 519). Clearly, the Chinese made their presence felt. For Rotblat, its four-man delegation had been "almost deliberately provocative and entirely non-cooperative."[53] As Rotblat went on, "Even some of the Russians were shocked, although Topchiev used all his influence to prevent the appearance of any divergence between Russian and Chinese policies." Rabinowitch likewise registered the Chinese problem, with one flashpoint being the Vienna Declaration, which they had refused to endorse because they had not been involved in its formulation, because of its assertions of common interest to all nations, and because they deeply resented its reference to "helping underdeveloped nations." In Rabinowitch's view, "their thesis is that each nation should follow the Chinese example of lifting itself by its own bootstraps, without asking for help—particularly not from the capitalist countries."[54] Rabinowitch reported that only after Topchiev had "spent hours trying to convince them" did the Chinese eventually commit to the Declaration and its principles. None of this augered well for Chinese relations with Pugwash. Indeed, after the Moscow conference, and in a major blow to Pugwash, the Chinese did not attend another conference until 1985.[55]

2.7 A Shift in the Official British View: Reassessing the PCSWA

Moscow proved decisive in bringing about a change in the attitude of the British government towards the PCSWA. Here, Alistair Buchan's report was important. Buchan filed copies of his report with Whitehall which carried news of the Soviet-American conversation into British government circles where his account of Moscow had powerful ripple effects. After the Second World War, the dilemmas of conscience arising from the use of the atomic bomb by the US in Japan in August 1945 led some senior scientists in the UK to walk away from nuclear weapons related research and, for some, including Joseph Rotblat, to actively protest against government policy in relation to these weapons. By contrast, others, including John Douglas Cockcroft (1897–1967) and William Penney (1909–1971), became leaders of the British nuclear program, becoming part of the nuclear establishment. These different

---

[53] Joseph Rotblat. 1961. Comments on Buchan's notes on the PCSWA in Moscow. RTBT 5/2/1/6 (38), 5–6.

[54] Eugene Rabinowitch. 1961. Thoughts on the meeting in Moscow. Draft Report, January 1961. RTBT 5/2/1/6 (39), 4–5.

[55] The People's Republic of China became a nuclear power on 16 October 1964, and a thermonuclear power on 17 June 1967.

responses to the atomic age created a faultline within British science between political insiders and outsiders. Yet, relationships between scientists across this divide
typically remained cordial, and scientific networks, even friendships, transcended
these differences. After all, elite scientists were colleagues in a broader, professional
sense, part of the same social and research networks of science, socializing at scientific conferences, reading each other's scientific papers, and sharing experiences of
laboratory life. Here was the 'scientific community' valorized in the Pugwash narrative and of which its scientists were a part. For the early Pugwash circle in Britain—a
British Pugwash group was not formally established until December 1962—scientific colleagues close to government formed a crucial bridgehead to senior government circles. This dynamic became more apparent after the Moscow conference as,
although still uneasy about the organization, the UK government began to perceive
it as relevant to its interests. Important here was Solly Zuckerman (1904–1993) at
this time based at the Ministry of Defence (MoD) and chief scientific advisor to the
British government (Dainton 1995; Krohn 1995, 575–598; Zuckerman 1975, 1980).

A zoologist specializing in primatology, Solly Zuckerman had risen rapidly within
the growing ranks of scientific advisors after the Second World War when invest-
ment in science rocketed globally and as the British government turned increasingly
to scientists for advice on science policy (Greenwood 1971a, b; Mencher 1969;
Gilpin 1970). South African by birth but based in Britain for almost his entire career,
Zuckerman was skilled at navigating the intersection between science and politics
and at straddling divides within science. During the war, along with John Desmond
Bernal (1901–1971) and others, he was appointed scientific advisor to the British
government by Louis Mountbatten. After the war, when Mountbatten was appointed
Chief of the Defence Staff, Zuckerman served him in an advisory role, during which
time the two men forged a close friendship. Subsequently, this friendship—dubbed
the "Zuck-Batten" axis—protected Zuckerman when, as was sometimes the case, he
expressed trenchant criticisms of colleagues and government policy (Maguire 2007;
Spinardi 1997). In the early 1950s, whilst building up the Anatomy Department at
Birmingham University, Zuckerman undertook science advisory work for successive Labour and Conservative governments, including in the early 1950s a stint as
UK representative on the Science Committee of NATO.[56] By 1960, Zuckerman was
Chief Scientific Advisor to the Ministry of Defence, and the country's most highly
regarded and influential scientific advisor to government. Yet Zuckerman also shared
a common history with those on the more radical left of British science. In the 1930s
he had mixed with leftist scientists including Bernal and Patrick M. S. Blackett
(1897–1974) through the *Tots and Quots* dining club founded by Zuckerman in 1930
(Green 1978; Werskey 1971).[57] At the outbreak of war, *Tots and Quots* "served as an
informal link between US, French and British scientists who had become involved in

---

[56] Between 1960 and 1963 Zuckerman was serving on the country's Defence Research Policy
Committee and he also served on the Committee which was reviewing the use and development
of scientific resources, especially manpower and education: the resulting ('Barlow') report of 1963
transformed the landscape of the British University system.

[57] *Tot homines, quot sententiae* (As many men, so many options). Membership of this informal
club included twenty or so natural scientists, sociologists, and economists: its monthly dinners

the war effort" (Zuckerman 1975, 467; Agar and Balmer 1998). Zuckerman was thus no stranger to moving in both dissenting and establishment circles, and accustomed to working informally across national borders (Rindzeviciute 2016). He also proved adept at straddling the divides within British science during the interwar period, in the early nuclear age and during the Cold War.

By early January 1961, Zuckerman had received Alistair Buchan's report on the Moscow conference which he took very seriously.[58] As a first step, he sent a copy to Rotblat inviting him to give his own views on the conference—an indication of how scientists negotiated and worked across their different standing in government circles.[59] Rotblat's reply was mainly a point-by-point response to Buchan. He began by briefly highlighting some of the positives he had taken from Moscow, moving also into ambassadorial mode for the PCSWA, taking every opportunity to advertise its role as a site of cross-bloc exchange and emphasizing how, in contrast to Buchan, his analysis was "based on the more frequent contacts" with the Soviets which, as he emphasized, helped to understand their standpoint.[60]

In response to Buchan's question as to "whether we were talking to the right people in Moscow," Rotblat acknowledged that "It may be a long time before we are able to answer" this important point, but countered that "there can be little doubt that some at least of the Soviet participants carry considerable weight with their government." In Rotblat's view, it was unlikely in the "highly centralized" Soviet system, that the "executive leaders of the powerful Academy of Sciences would not be people who are trusted and whose opinion is highly rated in official circles." Echoing Young's view, Rotblat perceived that Federov was "taking over real command," although Topchiev remained the official leader of Soviet Pugwash.[61] This was part of a wider point which Rotblat made to Zuckerman, that "We usually think of Soviet scientists in terms of a monolith, all voicing the official policy. This may not be correct," adding that he had noticed amongst the Soviets "a divergence of opinion on disarmament problems." Moreover, he urged the need for cultural sensitivity as a means to understand the different approach of the Soviets, which was "based on an entirely different ideology, tradition, mentality."[62] Agreeing with Buchan about the importance of private conversations, Rotblat again sought to advertise Pugwash strengths, emphasizing the rich opportunities for informal, ad hoc and private exchanges at the conference, but also during the busy social program. Conceding that "both sides misbehaved on a few occasions"—Topchiev's opening outburst being a case in point on the part of the Soviets—Rotblat felt both sides "gave on the whole an example of patience, restraint

brought guest speakers from within the political, military and scientific establishments, including those prominent in public affairs.

[58] Alistair Buchan. 1960a. Notes on Pugwash, Moscow, 27.11.1960–5.12.1960. RTBT 5/2/1/6 (39).

[59] Solly Zuckerman to Joseph Rotblat, 16 January 1961. RTBT 5/2/1/6 (38). Joseph Rotblat. 1961. *Comments on Buchan's notes on the PCSWA in Moscow.* RTBT 5/2/1/6 (38).

[60] Joseph Rotblat to Solly Zuckerman, 23 January 1961. RTBT 5/2/1/6 (38).

[61] Joseph Rotblat. 1961. *Comments on Buchan's notes on the PCSWA in Moscow.* RTBT 5/2/1/6 (38), 1.

[62] Ibid, 4.

and a genuine desire to understand each other's views."[63] In a rejoinder to Buchan's assessment that little "headway" had been made in Moscow, Rotblat argued that actually there had been some "very profitable discussions." Rotblat's reply reveals him to have developed into a nuanced operator as effective when dealing with the UK establishment as he was in the realm of informal diplomacy on the international stage of the Pugwash conferences.

Rotblat had honed these skills through his role as Secretary General—a role which, in an ad hoc, contingent way he largely defined, and the hallmark of which was impartiality—a crucial asset when it came to balancing East–West interests within the PCSWA. When dealing with his Soviet and also Eastern European colleagues, his Polish roots were undoubtedly an asset, but he was equally attuned to the political and cultural sensibilities of his American colleagues. This was crucial since the leadership was always actively navigating a path between the superpowers and had at all times to manage carefully its relations with Moscow and Washington. Linked to this, the Continuing Committee had always to strike a balance between US and Soviet influence within the organization. Senior figures in the American and Soviet Pugwash groups were keenly aware of this and looked here to their British colleagues. As the Pugwash organization developed, patrolling and maintaining this balance often fell to Rotblat, not least in his role as Secretary General, and to his British colleagues on the Continuing Committee, Neville F. Mott (1905–1996), Cecil F. Powell and Rudolf Peierls. Indeed, Rotblat hinted at this in his reply to Zuckerman about Buchan's report on the Moscow conference, making the point that although the British participants were numerically outweighed by the Americans and Soviets, their role in the discussions "carried considerable weight. The Russians listen to the British with much less suspicion than to the Americans. The British team can play a very important part in influencing Soviet opinion"—adding the same was true for the Chinese. Of course, in part, this was strategic—emphasizing to Zuckerman the important role of the British in the PCSWA, but it also reflected the realities of the internal dynamics of the organization. This aspect of the British role was, in practice, crucial to enabling the organization to operate at and across the East–West divide.

Zuckerman planned next to discuss Moscow with Jerome Wiesner who, with Kennedy now installed in the White House, he regarded as his opposite number in Washington.[64] In Whitehall, Buchan's report had cast light on the relationships and understanding that western scientists and their Soviet colleagues built within Pugwash, especially perhaps within the Continuing Committee. Perhaps there was particular concern about the extent of the Soviet-American dialogue in Moscow, including plans for the bilateral disarmament study group that became SADS. Perhaps Zuckerman had gleaned a sense of considerable British influence within the senior echelons of the PCSWA. The real question facing Zuckerman and his Whitehall colleagues was whether the British government could afford not to be present alongside the Americans and the Soviets at the Pugwash table. Zuckerman now perceived that Pugwash conferences had become generally useful as a place for

---

[63] Ibid, 4–5.

[64] Solly Zuckerman to Joseph Rotblat, 16 and 24 January 1961. RTBT 5/2/1/6 (38).

ad hoc discussions, including exploratory, "non-committal" diplomacy on issues that mattered to governments—which had engendered a recalibration of his assessment of it. What is clear is that reading Buchan's account of the Moscow conference stirred up Zuckerman who spread news of it to his colleagues in Whitehall.

The changing attitude of the British Foreign Office and the Ministry of Defence towards the PCSWA was apparent in the run up to the next conference in the US, scheduled for September 1961. A flurry of correspondence across the summer of 1961 between Zuckerman, British nuclear supremo John Cockcroft, and Foreign Office officials including A. Duncan Wilson reveal their attempts to rid British Pugwash of its dissenting elements so as to render it more compliant, and bring it more into line with and better serve "official" UK interests (Hartcup and Allibone 1984, 113–179; Rubinson 2011). Bertrand Russell was an early target. In June 1961, Wilson informed Zuckerman that Wayland Young "was aware that Lord Russell's connexion (to the Pugwash Conferences) was now an embarrassment, but thought that some of the British members, including Joseph Rotblat, might not wish to drop him, especially as he was no longer playing an active part. We have informally suggested that it might be wise to try to find a way round this difficulty soon. After all, the Americans seem to have disposed of Cyrus Eaton without much difficulty" (Sachse 2020).[65] A few days later, Cockcroft wrote to Wilson's colleague at the FO, Hugh Stephenson, suggesting that, as had been the case in Moscow and was the case for the upcoming American Conferences, which had been sponsored by the respective national academies of science, the Royal Society might sponsor "further British participation."[66] Cockcroft mentioned that he had already "had a brief word" about this with Royal Society President, Sir Howard Florey, who was "not averse to this." Cockcroft then added that he saw in this a means for the "winding up of the present Pugwash Committee," that is so say, replacing it with scientists from within the Royal Society fold who would follow the "Establishment" line. On 12 July, Wilson wrote to Cockcroft, stating that it would be "most helpful" if the Royal Society were to do this, and repeated that this might be a way "to lead to the winding up of the present (British) Pugwash Committee" and "would be a convenient way of getting over the problem" of Lord Bertrand Russell.[67] Interestingly, Wilson went on to suggest that the Royal Society retain the involvement of some of the existing members of the British Group, notably Wayland Young. This suggests that Young was considered the more acceptable face of British Pugwash, presumably because he would be more amenable to the Foreign Office. This raises questions about Young's position within Pugwash. Certainly, there were simmering tensions between him and Rotblat on matters relating to publications and the disclosure of Pugwash business. Whether

---

[65] A. Duncan Wilson to Solly Zuckerman, 16 June 1961. AB 16/2478, AEA Records.

[66] John Cockcroft to Hugh Stephenson [Foreign Office], 26 June 1961. AB 16/2748, Atomic Energy Authority (AEA) Records. In: Notes by Lawrence Wittner on British Government Records at the British National Archives, Kew, London. RTBT 5/1/3/7. The Continuing Committee had decided that the next conference in 1962 would be in the UK, plans for which were already underway.

[67] A. Duncan Wilson to John Cockcroft, 12 July 1961, AB 16/2748, AEA Records. RTBT 5/1/3/7.

news of the Royal Society machinations reached Rotblat remains unclear, but worsening relations between the two men proved irreparable and culminated in Young's departure from Pugwash in August 1962.[68]

In 1962, an internal Foreign Office memo written by Wilson noted that until 1960, the PCSWA had not been "considered politically very 'respectable' because there was a danger that they would simply become another Communist front organization or, at best, from the British point of view, one in which unilateralist views would be encouraged. These less respectable tendencies were represented both in the international side of the movement and in the British Committee."[69] The less "respectable tendencies" in Britain were likely a reference to Russell, Rotblat, Cecil Powell and Eric Burhop, and internationally to Americans such as Cyrus Eaton and Linus Pauling (Frank and Perkins 1971; Massey and Davis 1981; Sachse 2020).[70] But Wilson noted that in 1960 there had been "a successful effort from the British and American side to put fellow travelers into eclipse and replace them with people of genuinely independent views." Although characterizing the shift in terms of successfully sidelining "fellow travelers" is not quite how Rotblat and Rabinowitch might have described the changes, efforts were underway at this time within Pugwash to distance the organization from those perceived to be attracting controversy and/or damaging its reputation. Carola Sachse has recently examined the case of Cyrus Eaton, and it is clear that Russell was of concern even in senior western circles within Pugwash—but Rotblat in particular was reluctant to force Russell out (Sachse 2020). In any case, Rotblat and his British Pugwash colleagues would also have seen the reference to those holding "genuinely independent views" for what it was: those whose links to the political establishment meant that they viewed the PCSWA through the prism of government interests.

Foreign Office assessments of the PCSWA and sustained discussions about intervening in its internal affairs so as to exercise stronger influence over it can be interpreted as recognition within the British government of its potential usefulness not least as a forum for cross-bloc dialogue and, following from this, for informal diplomacy. By 1963, Foreign Office concerns had shifted to Rotblat who, as A.D.F. Pemberton-Piggott lamented, "seems to be in sole charge" of British Pugwash and to the chagrin of the Foreign Office was fiercely protective of "his independence and scientific integrity," making it impossible to get him to "pay any attention to what we think."[71] Pemberton-Piggott went on to complain that for the upcoming 1963 Conference, held in Dubrovnik, "neither the Foreign Office nor Solly Zuckerman's office were consulted about the composition of the British delegation or what they were to say." The following year, another Foreign Office staffer, B.T. Price, reporting on

---

[68] Minutes of the fifteenth meeting of the Continuing Committee, August 1962. RTBT 5/3/1/2 (Pt. 2).

[69] A. D. Wilson to C. J. Hayes. Memo dated 26 April 1962. FO 371/163160. RTBT 5/1/3/7.

[70] Cecil Powell and Eric Burhop played extremely important roles in the inception of the PCSWA, with Powell—who simultaneously was a leading figure in the World Federation of Scientific Workers—remaining highly influential within Pugwash until his sudden and untimely death in 1969.

[71] A. D. F. Pemberton-Piggott, Foreign Office Minute, 25 October 1963. RTBT 5/1/3/7.

the Dubrovnik conference, begrudgingly observed that it is "in the corridors that the real argument goes on" and inferred that Pugwash scientists were less well equipped for these encounters than trained Foreign Office staff and/or diplomats. He went on to assert that "the price to be paid" for these opportunities—which, tellingly, he conceded "after all occurs nowhere else"—was a "good deal of tedium in the working sessions;" it was, he said, "a kind of diplomatic pearl-fishing."[72] Price continued, "and yet even when fishing a little planning is no bad thing. The Russians carefully interload their own scientists with influential political, diplomatic and military observers. Perhaps if we followed their example, we might in time manage to land more pearls."

That the British political establishment was seeking to seed Pugwash with its scientists in order to 'land more pearls' was apparent at the eighth conference in Stowe where the UK participants included John Cockcroft and Solly Zuckerman, fresh from their machinations as to how to effect change within the British group, and William Penney, director of the UK Atomic Energy Research Establishment at Harwell who had longstanding ties to government. Stowe afforded Zuckerman a first opportunity to talk with "Russian government scientists face-to-face" (Hartcup and Allibone 1984, 274). In the run up to Stowe, British participants held a series of preparatory meetings, including at Caius College, Cambridge. Tellingly, they were also called to the Foreign Office on 28 August to meet with Duncan Wilson, a meeting coordinated by John Cockcroft.[73] Zuckerman knew Cockcroft from their days together serving on the Advisory Council on Scientific Policy during the 1950s (Hartcup and Allibone 1984, 113–179).[74] This would be the first Pugwash Conference for both Cockcroft and Zuckerman; Cockcroft already knew some of the Soviet Pugwash scientists from his work in the international nuclear realm, including Vasily S. Emelyanov alongside whom, since 1958, he worked on the Scientific Advisory Committee of the IAEA, and Dimitrii Skobel'tsyn from their days serving on the UN Scientific Advisory Committee, established by Dag Hammerskjöld early in 1955 (Hartcup and Allibone 1984, 233). This illustrates how, by virtue of their seniority, scientists got to know each other in the course of international roles and duties, activities which rendered them part of what could be described as a global—if loose—scientific network (Wunderle 2015). In effect, in seeking the participation of senior scientists around the world, the PCSWA was tapping into pre-existing scientific networks established not only through shared research interests, but by shared duties serving on various national and international committees, boards, and panels, that were now more than ever a routine feature in the busy diaries of senior scientists everywhere.

---

[72] B. T. Price. Report on Dubrovnik Conference, 09/1963. 16 October 1963. Marked confidential. FO 371/171/90. RTBT 5/1/3/7.

[73] Joseph Rotblat to Patrick M. S. Blackett, 10 August 1961. P. M. S. Blackett to Joseph Rotblat, 14 August 1962. RTBT 5/2/1/7 (9). Philip Noel-Baker to Joseph Rotblat, 25 August 1961. RTBT 5/2/1/7 (9).

[74] At the Council, the Zuckerman and Cockcroft had, for example, worked on policy relating to British scientific education and manpower.

2.8 Moscow: Watershed and Platform

The Moscow conference brought about a quiet but profound transformation in the PCSWA. The American and the Soviet scientists had shared a considerable amount of information—conceptual, theoretical, technical—during nine days of discussions on a range of disarmament problems. In the course of these exchanges, these scientists gained an understanding of each other and of the national mindset. This had placed relations between them on a new footing with the two sides making headway in establishing a *modus vivendi* for working together within the Pugwash framework. At the conference in Baden a year earlier, the statement had included the observation that "Trust between nations cannot be established by proclamation, but only by experience, particularly by experience in co-operative work towards common aims" (Chisholm et al. 1959, 1020).[75] Something along the lines of "co-operative work" had taken place in Moscow. This was unprecedented, and if these relationships remained cautious and tentative, the American and Soviet scientists had agreed to move forward with the idea for creating the East–West Study group on disarmament. As Bernd Kubbig has shown, this group—which from the outset had close links with the White House and with the Kremlin—came later to play a decisive role in the complex process leading to the completion of the Anti-Ballistic Missile Treaty in 1972 (Kubbig 1996).

It is clear too that in Moscow, the informal *modus operandi* of the Pugwash organization was beginning to pay dividends, not least in creating unparalleled opportunities for those attending its conferences to hold private, informal meetings, and exploratory exchanges with colleagues from both the same and the opposite side of the bloc divide (Kubbig 1996, 8, 13). This suddenly registered the conferences anew with state actors, including the British, which now perceived that the PCSWA might be useful to them as a forum for unofficial informal diplomacy. At the very least, Whitehall was moving to a view that it could not afford to miss out on what went on at Pugwash conferences.

In hindsight, the Moscow conference was a turning point for the Pugwash organization in terms both of its status with the American, Soviet and British governments and its emerging role in informal back-channel diplomacy between the superpowers. But in terms of China, Moscow brought disappointment. The difficulties with the Chinese foreshadowed the country's twenty-five year absence from the PCSWA. In terms of its internal development, the procedural innovation of the break-out groups had far-reaching importance for Pugwash, providing the spur to the introduction in Stowe of Working Groups—which, henceforth, became a mainstay of the work carried out at the conferences.

---

[75] Statement from the fifth PCSWA, Nova Scotia, 24–29 August, 1959.

References

1960. Table summarizing Soviet and Western proposals on disarmament. Unattributed. *Bulletin of the Atomic Scientists* 16 (8): 336–339.

Agar, Jon, and Brian Balmer. 1998. British scientists and the Cold War: The Defence Research Policy Committee and information networks, 1947–1963. *Historical Studies in the Physical and Biological Sciences* 28 (2): 209–252.

Barrett, Gordon. 2020. Minding the gap: Zhou Peiyuan, Dorothy Hodgkin, and the durability of Sino-Pugwash networks. In *Science, (anti-)communism and diplomacy: The Pugwash Conferences on Science and World Affairs in the early Cold War,* eds. Alison Kraft, and Carola Sachse, 190–217. Leiden: Brill.

Bergel, Franz. 1977. Alexander Haddow. *Biographical Memoirs of Fellows of the Royal Society* 23: 132–191.

Blagonravov, Anatoli A. 1960. Control over means of delivery. In *Proceedings of the 6th Pugwash Conference, Moscow November 1960*, 573–579.

Brennan, Donald. 1960. On flexibility, communication and a specific arms proposal. In *Proceedings of the 6th Pugwash Conference, Moscow November 1960*, 608–628.

Buchan, Alistair. 1960. Surprise attack. In *Proceedings of the 6th Pugwash Conference, Moscow November 1960*, 542–553.

Chisholm, Brock, Claude E. Dolman, Donald Kerr, Robert Watson-Watt, Preben Von Magnus, Andre Lwoff, Pierre Thibault, M.L. Ahuja, Mikhail M. Dubinin, Alexandre A. Imshenetsky, Vladimir P. Pavlichenko, A.A. Smorodintsev, Sven Gard, F.C. Bawden, Patricia J. Lindop, Gordon Manley, Joseph Rotblat, M.G.P. Stoker, H. Bentley Glass, Charles C. Higgins, Martin M. Kaplan, Chauncey D. Leake, Hugo Muench, Eugene Rabinowitch, Alexander Rich, and Theodor Rosebury. 1959. Pugwash International Conference of Scientists: Statement on biological and chemical warfare. *Nature* 184: 1018–1020.

Daedalus. 1960. Special issue on arms control. 89 (4).

Dainton, Frederick S. 1995. Lord Zuckerman. *Proceedings of the American Philosophical Society* 139 (2): 212–217.

Damms, Richard V. 2000. Jack Kennedy, the technological capabilities panel, and the emergence of President Eisenhower's scientific-technological elite. *Diplomatic History* 24 (1): 57–78.

Dodd, Thomas J. 1961. The Pugwash conferences: A staff analysis. In *Internal Security Subcommittee, 87th Congress, 1st Session 1. US Congressional Record*, vol. 107, Pt. II, 15059. Washington: Government Printing Office.

Doty, Paul. 1960. Current attitudes on disarmament in America. In *Proceedings of the 6th Pugwash Conference, Moscow November 1960*, 35–42.

Dubinin, Michail. 1960. Modern chemical weapons are weapons of mass destruction. In *Proceedings of the 6th Pugwash Conference, Moscow November 1960*, 451–555.

Edsall, John T. 1960. Minutes of thirteenth session. In *Proceedings of the 6th Pugwash Conference, Moscow November 1960*, 791.

Edwards, A.P.J. 1994. Ben Lockspeiser. *Biographical Memoirs of Fellows of the Royal Society* 39: 246–261.

Evangelista, Matthew. 1999. *Unarmed forces. The transnational movement to end the Cold War.* Ithaca, New York: Cornell University Press.

Federov, Evgenii, K. 1960. The present stage of talks on termination of nuclear tests. In *Proceedings of the 6th Pugwash Conference, Moscow November 1960*, 187–197.

Feld, Bernard T. 1960. Inspection problems of arms control and disarmament. In *Proceedings of the 6th Pugwash Conference, Moscow November 1960*, 398–408.

Frank, F.C., and D.H. Perkins. 1971. Cecil F. Powell, 1903–1969. *Biographical Memoirs of Fellows of the Royal Society* 17: 541–563.

Garwin, Paul, and E. Peter Geiduschek. (2013). Paul M. Doty 1920–2011. *Biographical Memoirs.* National Academy of Sciences: 1–27.

Geiger, Roger L. 1997. What happened after Sputnik? Shaping university research in the US. *Minerva* 35: 349–367.

Gilpin, Robert. 1970. Technological strategies and national purpose. *Science* 169: 441–448.

Green, Martin. 1978. The visible college in British science. *The American Scholar* 47 (1): 105–117.

Greenwood, John W. 1971a. The scientist-diplomat: A new hybrid role in foreign affairs. *Science Forum* 19: 14–18.

Greenwood, John W. 1971b. The science attaché: Who he is and what he does. *Science Forum* 20: 21–25.

Hartcup, Guy, and Thomas E. Allibone. 1984. *Cockcroft and the atom.* Bristol: Adam Hilger.

Higuchi, Toshihiro. 2020. *Political fallout. Nuclear weapons testing and the making of a global environmental crisis.* Stanford, CA: Stanford University Press.

Howard, Michael. 2004. *Buchan, Alastair Francis.* Oxford Dictionary of National Biography. https://doi.org/10.1093/ref:odnb/30868. Accessed 25 Aug 2021.

Imshenetski, Aleksandr D. 1960. Ban biological weapons. In *Proceedings of the 6th Pugwash Conference, Moscow November 1960.*

Kaplan, Martin M. 1999. The efforts of WHO and Pugwash to eliminate chemical and biological weapons—A memoir. *Bulletin of the World Health Organization* 77 (2): 149–155.

Kraft, Alison. 2018. Dissenting scientists in early Cold War Britain. The "Fallout" controversy and the Origins of Pugwash, 1954–1957. *Journal of Cold War Studies (JCWS)* 20 (1): 58–100.

Kraft, Alison, and Carola Sachse. 2020. The Pugwash Conferences on Science and World Affairs: Vision, rhetoric, realities. In *Science, (anti-)communism and diplomacy: The Pugwash Conferences on Science and World Affairs in the early Cold War,* eds. Alison Kraft, and Carola Sachse, 1–39. Leiden: Brill.

Krohn, P.L. 1995. Solly Zuckerman. Baron Zuckerman, of Burnham Thorpe, O.M., K.C.B. *Biographical Memoirs of Fellows of The Royal Society* 575–598.

Kubbig, Bernd W. 1996. *Communicators in the Cold War: The Pugwash Conferences, The U.S.-Soviet Study Group and the ABM treaty. Natural scientists as political actors: Historical successes and lessons for the future.* PRIF Reports No. 44. Frankfurt am Main: PRIF.

Laucht, Christoph. 2012. *Elemental Germans. Klaus Fuchs, Rudolf Peierls and the making of British nuclear culture 1939–1959.* Basingstoke, UK: Palgrave Macmillan.

Lüscher, Fabian. 2020. Party, peers, publicity. In *Science, (anti-)communism and diplomacy: The Pugwash Conferences on Science and World Affairs in the early Cold War,* eds. Alison Kraft, and Carola Sachse, 121–155. Leiden: Brill.

Lüscher, Fabian. 2021. *Nuklearer Internationalismus in der Sowjetunion. Geteiltes Wissen in einer geteilten Welt 1945–1973.* Böhlau: Wien.

Macdonald, Julia M. 2015. Eisenhower's scientists: Policy entrepreneurs and the test-ban debate 1954–1958. *Foreign Policy Analysis* 11: 1–21.

Maguire, Richard. 2007. Scientists dissent amid the British Government's nuclear weapons program. *History Workshop Journal* 63 (1): 113–135.

Manzione, Joseph. 2000. "Amusing, and amazing and practical and military:" The legacy of scientific internationalism in US foreign policy, 1945–1963. *Diplomatic History* 24 (1): 21–55.

Massey, Harrie, and D.H. Davis. 1981. Eric Henry Stoneley Burhop. *Biographical Memoirs of Fellows of The Royal Society* 27: 131–152.

Mencher, A.G. 1969. Scientists among diplomats. *Bulletin of the Atomic Scientists* 46–48.

Pasqualucci, J.M. 1998. Louis Sohn. *Human Rights Quarterly* 20 (4): 924–944.

Pei-Yuan, Zhou. 1960. The disarmament problem and the points of view of the Chinese scientists. In *Proceedings of the 6th Pugwash Conference, Moscow November 1960*, 515–522.

Perry Robinson, J.P. 1998. The impact of Pugwash on the debates over chemical and biological weapons. *Annals of the New York Academy of Sciences* 866 (1): 224–252.

Rabinowitch, Eugene. 1960. Creation of a suitable climate for disarmament. In *Proceedings of the 6th Pugwash Conference, Moscow November 1960*, 688–708.

Rabinowitch, Eugene. 1965. About Pugwash. *Bulletin of the Atomic Scientists* 21 (4): 9–15.

Rindzeviciute, Egle. 2016. *The power of systems. How policy sciences opened up the Cold War world.* Ithaca: Cornell University Press.

Rosenblith, Walter A. 2003. *Jerry Wiesner: Scientist, statesman, humanist: Memories and memoirs.* Cambridge, MA: Massachusetts Institute of Technology Press.

Rotblat, Joseph. 1960. *Proceedings of the 6th Pugwash Conference, Moscow, 1960,* 292.

Rotblat, Joseph. 1967. *Pugwash—A history of the Conferences on Science and World Affairs.* Prague: Czechoslovak Academy of Sciences.

Rotblat, Joseph. 1998. A social conscience for the nuclear age. In *Hiroshima's shadow,* eds. Kai Bird, and Lawrence Lifschultz, xvii–xxviii. Stony Creek: Pamphleteer's Press.

Rubinson, Paul. 2011. "Crucified on a cross of atoms." Scientists, politics and the Test Ban Treaty. *Diplomatic History* 35 (2): 283–319.

Rubinson, Paul. 2016. *Redefining science: Scientists, the national security state and nuclear weapons in Cold War America.* Amherst, Boston: University of Massachusetts Press.

Rubinson, Paul. 2020. American scientists in "Communist Conclaves." Pugwash and anti-communism in the US. In *Science, (anti-)communism and diplomacy: The Pugwash Conferences on Science and World Affairs in the early Cold War,* eds. Alison Kraft, and Carola Sachse, 156–189. Leiden: Brill.

Sachse, Carola. 2020. Patronage impossible: Cyrus Eaton and his Pugwash scientists. In *Science, (anti-)communism and diplomacy: The Pugwash Conferences on Science and World Affairs in the early Cold War,* eds. Alison Kraft, and Carola Sachse, 80–117. Leiden: Brill.

Sohn, Louis B., and David H. Frisch. 1960. Arms control in the 1960s. In *Proceedings of the 6th Pugwash Conference, Moscow November 1960,* 103–125.

Spinardi, Graham. 1997. Aldermaston and British nuclear weapons development: Testing the "Zuckerman thesis." *Social Studies of Science* 27: 547–582.

Talensky, Nikolai A. 1960. The technical problems of reducing armaments. In *Proceedings of the 6th Pugwash Conference, Moscow November 1960,* 409–417.

The Nobel Peace Prize. 1959. NobelPrize.org. Nobel Prize Outreach AB 2021. https://www.nobelprize.org/prizes/peace/1959/summary/. Accessed 25 Aug 2021.

Topchiev, Aleksandr. 1960. Disarmament: The most urgent problem of today. In *Proceedings of the 6th Pugwash Conference, Moscow November 1960,* 21–32. Copy also in RTBT 5/2/1/6 (7).

Vinogradov, Alexander P. 1959. Prospects for the Pugwash movement. *Bulletin of the Atomic Scientists* 15: 376–378.

Werskey, Gary. 1971. British scientists and 'outsider' politics, 1931–1945. *Science Studies* 1: 67–83.

White, Theodore H. 1967. The action intellectuals. *Life Magazine.*

Wiesner, Jerome. 1960. Comprehensive arms limitations systems. In *Proceedings of the 6th Pugwash Conference, Moscow November 1960.*

Wiesner, Jerome. 1963. John F. Kennedy: A remembrance. *Science* 142 (3596): 1147–1150.

Wunderle, Ulrike. 2015. *Experten im Kalten Krieg: Kriegserfahrungen und Friedenskonzeptionen US-amerikanischer Kernphysiker 1920–1963.* Leiden: Brill.

Zuckerman, Solly. 1975. Scientific advice during and since WWII. *Proceedings of the American Philosophical Society* 342: 465–480.

Zuckerman, Solly. 1980. Science advisers and scientific advisers. *Proceedings of the American Philosophical Society* 124 (4): 241–255.

Chapter 3
Stowe, Vermont, September 1961: Two Contrasting Conferences in America

Abstract The focus in this chapter crosses the Atlantic to Stowe, Vermont, where the seventh and eighth Pugwash conferences were held in September 1961. For the first time, the conferences included Working Groups, marking a hugely successful innovation. The seventh conference focused on how Pugwash could foster international scientific cooperation, including across the North–South divide. Ideas were optimistically floated for the creation of large international laboratories, including notably in Berlin, and for a WHO-sponsored World Health Research Center. By contrast, the eighth conference, dedicated to disarmament, was poisoned by the Berlin crisis and the recent resumption of nuclear tests by the USSR, triggering acute tensions between the superpowers. This political crisis made for Cold War grandstanding and hostilities in Stowe, exposing the myth that scientists could suspend national and bloc loyalties. Rather, Pugwash could never fully escape the divides it sought to transcend. Nevertheless, amid all the acrimony, the organization survived and the dialogue on disarmament continued.

Keywords Working Groups · The Soviet-American Disarmament Study Group (SADS) · Evgenii Federov · Martin M. Kaplan: Henry Kissinger: Jerome Wiesner · Solly Zuckerman · Sub-governmental diplomacy

3.1 A Change of Conference Plans, a Downturn in the Superpower Relationship

In early 1961, hopes were high within Pugwash that in America the organization could build on the platform of understanding which many thought had been established in Moscow—especially with John F. Kennedy newly installed in the White House. However, such hopes were dashed by sharply deteriorating relations between the superpowers during 1961. Indeed, that the conference scheduled for September in the US was not completely derailed was arguably significant in itself. That neither Moscow or Washington stood in the way of their scientists travelling to Vermont could be taken as an indication that both governments considered it worthwhile for the conference to go ahead. Perhaps they were keen to keep the bilateral disarmament

© The Author(s), under exclusive license to Springer Nature Switzerland AG 2022   45
A. Kraft, *From Dissent to Diplomacy: The Pugwash Project During the 1960s Cold War*,
SpringerBriefs in History of Science and Technology,
https://doi.org/10.1007/978-3-031-12135-7_3

study group conceived in Moscow afloat. But the changed political context poisoned the mood in Stowe—confirmation, if any were needed, of how Pugwash was powerfully shaped by the vagaries of the superpower relationship. Deep divisions along bloc lines were starkly apparent during the second conference—confirmation too that, for all the rhetoric to the contrary, Pugwash scientists, including those on the Continuing Committee, were not always able to suspend national allegiances.

In early 1961, the plans for the upcoming American conference changed. This followed a meeting of the American Pugwash group in January, where Eugene Rabinowitch argued that it should encompass the theme of international scientific cooperation—to which he had long been committed (Long 1972; Feld 1973; Slaney 2012). He saw such cooperation as a means to harness the internationalist tradition of science for peaceful goals, and there were recent precedents, notably the International Geophysical Year, the International Atomic Energy Agency and the CERN laboratory in Geneva—often held up as a resounding success and the exemplar for future European scientific cooperation in the 1960s Cold War (Aronova et al. 2010; Bulkeley 2000; Collis and Dodds 2008; Fischer 1997; Röhrlich 2018; Hermann et al. 1990; Krige and Guzzetti 1997; Krige and Barth 2006; Krige 2016; Krige and Oreskes 2014). Already at the Moscow conference, Rabinowitch had emphasized scientific cooperation as a powerful means to counter mistrust and to reduce international tensions, all of which could help create a "climate" conducive to disarmament. Moreover, he saw such cooperation as a means to build links with and strengthen the economies of the countries of the so-called "developing" world. For him, countries defined in this way had hitherto been too often regarded as a "side show" to the "main ring" of the superpower conflict (Rabinowitch 1965, 11–12). Rabinowitch saw cooperation here as both a humanitarian imperative and, in reducing north–south inequities, a powerful means to foster international stability and security—especially in a context in which the countries of the Global South were increasingly construed as 'proxies' for the superpower rivalry. For him too, this engagement provided a means for Pugwash scientists to put the principle of social responsibility into practice. His colleagues in the American group—Harrison Brown (1917–1986), Hiram Bentley Glass (1906–2005), Alexander Rich—took his point. That said, all understood that nuclear disarmament remained the Pugwash priority. As a compromise, the US Planning Committee proposed to hold two separate, consecutive conferences—a highly unusual arrangement that was approved by the Continuing Committee in March.[1] The first Stowe conference would focus on international scientific cooperation whilst the second returned to the familiar territory of disarmament and world security (Rotblat 1967, 29).[2] In the Pugwash chronology, these would be the seventh and eighth conferences and they took place between 5–9 September and 11–16 September respectively. The venue was a secluded retreat in Vermont called Smuggler's Notch.

---

[1] Minutes of the twelfth meeting of the Continuing Committee, 5–7 March 1961, London. RTBT 5/3/1/2 (Pt. 1) (3).

[2] Summary of the Stowe Conferences, Typed Manuscript, undated. RTBT 5/2/1/7 (12). Meeting of the Continuing Committee, March 1961. RTBT 5/3/1/2 (Pt. 2).

The Stowe conferences were the first international Pugwash events to take place in the US. They were hosted jointly by the National Academy of Sciences and the American Academy of Arts and Sciences (AAAS) and majority funded by the Ford Foundation. Organization was placed in the hands of a (US) Planning Committee chaired by Harrison Brown which, alongside Glass and Rabinowitch, included Richard Leghorn and Paul Doty, the latter serving as Vice-Chair.[3] The close involvement of Doty and Leghorn was a telling sign of the importance attached to the conferences and to the PCSWA at this moment in White House circles. Both were leading players within the American 'arms control' fraternity, and both had been in Moscow. At this time, Leghorn—a colleague of Alistair Buchan at the Institute for Strategic Studies in London—was involved in the disarmament talks in Geneva, whilst Doty was closely involved in the on-going discussions about creating the SADS group. Doty had contributed to the special issue of *Daedalus* in September 1960 which laid the foundations of the arms control approach to disarmament—on which, with Jerome Wiesner, he organized a panel at the celebrations marking the centennial anniversary of MIT in April 1961.[4] During the summer of 1961 when Brown spent a month in the Soviet Union, Doty took over his administrative role on the planning committee. The tally of East Coast luminaries involved in the preparations for Stowe also included Arthur Singer, the Dean of MIT, who was in overall administrative charge, with assistance from PCSWA veteran Ruth Adams who had a great deal of experience in running the conferences (Rubinson 2020).[5] As in Moscow, Wayland Young served as the Public Relations Officer for the European press, whilst the American press was handled by journalist and fundraiser for liberal social causes, Harold L. Oram.[6]

The deepening political crises during the summer of 1961—most spectacularly the building of the Berlin Wall in August—cast a "deep shadow" over Stowe. According to Rotblat, as early as July amid escalating tensions over Berlin, Harrison Brown had been "pessimistic" about the prospects for the conference. But Rotblat remained optimistic, as he put it, he "didn't believe that the Russians will want to break up Pugwash; in fact the worse the official situation the more anxious they will be to maintain unofficial contacts with the scientists."[7] This affords a telling insight into Rotblat's perception of the value attached to Pugwash by his Soviet colleagues. But the political situation deteriorated further when on 1st September, less than a week before Stowe was due to start, the Soviet Union carried out a nuclear weapons test, bringing to an abrupt end the moratorium in place since November 1958. This sent shockwaves through the Pugwash organization, and created tensions within the Continuing Committee.[8] The resumption of tests was a major setback to those

---

[3] Summary of Stowe Conferences, undated typed manuscript. RTBT 5/2/1/7 (12).

[4] MIT Centennial, Boston Sunday Globe, 2 April 1961.

[5] Harrison Brown to Harold Oram, 8 June 1961. RTBT 5/2/1/7 (11).

[6] Harrison Brown to Harold Oram, 8 June 1961. RTBT 5/2/1/7 (11). Oram was founder and president of the Oram group, a fundraising organization for humanitarian and social causes.

[7] Joseph Rotblat to Mary Morse, 20 July 1961. RTBT 5/2/1/7 (9).

[8] Minutes of thirteenth meeting of the Continuing Committee, Stowe, VT, 2–15 September 1961. RTBT 5/3/1/2 (Pt 1) (3).

working towards disarmament and against nuclear proliferation, and elicited great anger and dismay around the world. In response, in spring 1962 the US also resumed testing.

In this febrile atmosphere, the Stowe conferences went ahead. Attendance was again overwhelmingly dominated by the Americans, the Soviets and the British. Eugene Rabinowitch, Joseph Rotblat and Vladimir P. Pavlichenko were present and all three had also been at the Moscow conference (Rotblat 1967). Regular attendees steadily got to know each other, with the balance between respect and wariness differing from individual to individual. For example, the ever-present Pavlichenko was widely understood to be monitoring his Soviet colleagues and the conferences for the KGB: accordingly, he was viewed with unease and handled with care by the Continuing Committee (Kubbig 1996, 9; Lüscher 2020, 132). As in Moscow, those present included scientists moving in the corridors of power in East and West. The Soviet delegation at both Stowe conferences included seven scientists—Anatoli Blagonravov, Nikolai Bogoliubov, Mikhail Dubinin, Vladimir Khvostov, Nikolai Talensky, Igor Tamm and Aleksandr Topchiev—who had been at the conference in Moscow: in this period, this cohort formed the nucleus of Soviet Pugwash. In addition to Rabinowitch, the American party included Harrison Brown, Paul Doty, Bentley Glass and Jay Orear, all of whom had been in Moscow. Henry Kissinger (b. 1923) was at the eighth conference—marking his first appearance at the PCSWA. Notable by their absence at both Stowe conferences were Jerome Wiesner and Evgenii Federov, who had struck up a rapport in Moscow.

A year on from Moscow, these scientists were now coming together in a uniquely tense and hostile political context. The question as to how this would affect the dynamics in Stowe was uppermost in everyone's mind. In hindsight, the decision to focus the seventh conference on international scientific cooperation proved fortuitous. This theme was not politically controversial and agreement was easy to find, not least regarding the role that scientific cooperation could play in reducing tensions and building trust. At any rate, discussions flowed easily in an atmosphere of collegiality. The seventh PCSWA therefore functioned as something of an icebreaker for what proved to be an altogether more bitter and rancorous conference the following week.

3.2 The First Stowe Conference: An Icebreaker Amid Simmering Tensions

The seventh conference given to the theme 'International co-operation in pure and applied science' involved fifty participants, including twenty-four Americans, eleven Soviet scientists and six from Britain. One scientist each came from Australia, Austria, Brazil, Bulgaria, West Germany, Hungary, Italy, Japan and the Netherlands (Rotblat 1967, 121–123).[9] In his opening remarks, Harrison Brown, chair of the

---

[9] Specifically, for the US: twenty, plus four observers; for the USSR: eight, plus three observers; for the UK: four, plus two observers—Bullard, Haddow, Lockspeiser and Rotblat, with Patricia Lindop

US conference committee, acknowledged that they were meeting in "the middle of a serious international crisis," and that Pugwash was "perhaps the only significant channel of East–West communication open today."[10] Brown reminded his audience that the conference was private, with the press excluded—although daily briefings would be issued—and that Chatham House "rules" applied. The Continuing Committee was keen to control—as far as possible—how news of what went on reached the public domain.

Opening the first plenary session with a paper entitled 'International cooperation in science,' Eugene Rabinowitch likewise acknowledged that they were assembling under "dark and ominous clouds" amid the "thunderstorm gathering over Berlin," and with the new impetus given to the arms race "by the resumption of tests by the Soviet Union."[11] He emphasized that fostering international scientific cooperation was "only more important" in this context and, as he put it, the "best means I can think of to strengthen among nations the consciousness of their common interests, and to smooth the way to political reconciliation." This idealism was tempered by knowing that "we cannot work miracles, but we can do more than express pious wishes. We can attempt to develop specific proposals and outline concrete plans." Indeed, this was the Pugwash strategy for fostering international scientific cooperation, namely, scientists from East and West conceiving and developing outline plans for large scale supranational projects. These ideas would then be passed along—in an as yet unspecified process—to governments and/or international NGOs which had the requisite financial, material and human resources to take projects forward. There was no naiveté here: while proposed projects could potentially involve cooperation across the bloc divide, all recognized that any large international laboratories that would lie at the heart of such large supranational ventures presented complex political and logistical challenges, and could not be based in either the US or the Soviet Union.[12]

The conference program involved plenary sessions where prepared papers were discussed. Joseph Rotblat summarized Pierre Auger's recent book, *Current trends in scientific research*, written under the auspices of UNESCO (Rotblat 1961a, b). Aleksandr Topchiev and Roger Revelle offered, respectively, a Soviet and American perspective on the importance of international scientific cooperation (Topchiev 1961a; Revelle 1961). Notably, Topchiev's upbeat endorsement of such cooperation contrasted sharply with his defiant tone a week later. In a paper about international cooperation in cancer research, the British physician Alexander Haddow emphasized the need to build trust in order to move forward with disarmament—a point echoed by Dutch lawyer and peace researcher Bernard V. A. Röling (1906–1985) who lamented

---

and Wayland Young as "Observers." The category of "Observer" provided a degree of flexibility on conference attendance providing, on occasion, including at Stowe 7, a mechanism by which the Continuing Committee balanced out US and Soviet participation. Importantly, Observers were allowed to participate in the Working Groups.

[10] Harrison Brown. Notes on formal opening session. 5 September 1961. Seventh PCSWA, Stowe, VT. RTBT 5/2/1/7 (13).

[11] Eugene Rabinowitch. International cooperation in science. Paper VII–IV. Main program, PCSWA 7, Stowe, 5–9 September 1961. RTBT 5/2/1/7 (13), 2.

[12] The location of any such laboratories made for lengthy discussion.

the current "violent mistrust" between the superpowers (Haddow 1961; Röling 1961). In his lengthy paper, Haddow veered into sobering reflections on what he saw to be the pressing dilemma now confronting the PCSWA. Reflecting on the "sense of fuller optimism" in Moscow he felt that the organization was now entering "an intensely critical stage," as he put it:

> We have never minimized the immense inherent difficulties. We have never claimed the right of policy-making or decision, which is certainly not ours. Although I hesitate to say it, we must face the prospect of failure and defeat.

Haddow's narrative of fragility and failure preceded his call for the Pugwash organization to evolve so as to improve its ability to reach into and influence policy-making circles of national government (Kraft and Sachse 2020). Those present were all too aware that this remained the fundamental challenge facing the organization.

The program in Stowe featured an important innovation. Following the highly successful introduction of small break-out groups in Moscow, Working Groups were included for the first time as an integral part of the conference. A list of six Working Groups was issued at the registration desk, with colleagues invited to sign up to a maximum of two groups. The themes of the groups were: (I) Exchange of scientists and scientific information; (II) Cooperation in the earth sciences; (III) Cooperation in space research; (IV) Cooperation in the life sciences; (V) Cooperation in the physical sciences and (VI) Cooperation in assistance to developing nations[13] (Table 3.1). One full day was given to the Working Groups, followed by a plenary discussion of the draft reports from each group, with the final reports discussed in another plenary session. Henceforth, and at the discretion of the Continuing Committee, Working Group reports could be included in the Conference *Proceedings,* meaning they could reach a wider audience; they served also as a resource when it came to putting together the conference statement (Rotblat 1967). These Groups proved transformative to the quality and extent of the work carried out at the conferences, and in the next few years procedures in relation to them were devised in a pragmatic, 'learning by doing' process which started in Stowe.

With science and technology very much to the fore, discussions at the seventh conference steered clear of politically incendiary terrain, including within the Working Groups. The participants had been selected for their knowledge about both national science systems and experience in international science projects, including the political dimensions and the pitfalls of such collaboration. For example in a session of Working Group 5 discussing co-operation in the physical sciences, attention focused on the idea of building a new international particle accelerator perhaps

---

[13] *Proceedings of the 7th Pugwash Conference, Stowe, September 1961.* The chairs for the six Working Groups struck a careful balance between the US and USSR, as follows: I. Gerard Piel (US); II. Columbus O'Donnell Iselin (US); III. Anatoli A. Blagonravov (USSR); IV. Alexander Rich (US); Jerrold R. Zacharias (US) and Igor Tamm (USSR); VI. Edward Bullard (UK). Notes on 4th Plenary session, afternoon of 8 September 1961. RTBT 5/2/1/7 (13). It should be noted that occasionally there are discrepancies in the numbering of the groups as recorded in Rotblat's 1967 book on the PCSWA, in the *Proceedings,* and/or in records of the seventh PCSWA in Stowe in 1961 in the primary sources held in the Rotblat collection held at Churchill College, Cambridge, UK. Table 3.1 is based on the *Proceedings.*

Table 3.1 Working Groups: Seventh Pugwash Conference, Stowe, VT, USA, 5–9 September 1961[14]

Group	Topic	Members
1	Exchange of scientists and scientific information	**USSR**: A. A. Blagonravov, N. N. Bogoliubov, M. M. Dubinin, S. G. T. Korneev, V. P. Pavlichenko, I. E. Tamm **US**: D. Bronk, H. Brown, P. Doty, B. Glass, H. Hoagland, C. D. Leake, L. C. Mitchell, J. Orear, G. Piel, L. Pauling, A. M. Weinberg **UK**: A. Haddow **AUT**: H. Thirring **FRG**: G. Burkhardt **HUN**: F. B. Straub **Japan**: T. Toyodo
2	International cooperation in the earth sciences	**USSR**: A. Topchiev **US**: C. O'D Iselin, G. Piel, R. Revelle
3	International cooperation in space research	**USSR**: A. A. Blagonravov, N. N. Talensky **US**: H. Brown, W. H. Pickering, E. Rabinowitch, E. Wigner **UK**: E. Bullard **AUT**: H. Thirring
4	International cooperation in the life sciences	**USSR**: N. I. Sissakian **US**: D. Bronk, W. Conzolazio, P. Doty, B. Glass, H. Hoagland, M. M. Kaplan, C. D. Leake, A. Rich, E. Staley **UK**: A. Haddow, B. Lockspeiser **AUS**: J. Crawford **Brazil**: C. Pavan **HUN**: F. B. Straub
5	International cooperation in the physical sciences	**USSR**: N. N. Bogoliubov, M. M. Dubinin, V. P. Pavlichenko, I. E. Tamm **US**: J. Orear, L. Pauling, I. I. Rabi, A. M. Weinberg, J. R. Zacharias **UK**: J. Rotblat **BUL**: G. Nadjakov **FRG**: G. Burkhardt **Italy**: G. Bernardini **Japan**: T. Toyoda
6	Cooperation in assistance to developing nations	**USSR**: N. I. Bazanov, V. Khvostov, N. I. Sissakian, N. N. Talensky **US**: W. Consolazio, C. O'D Iselin, M. M. Kaplan, I. Rabi, E. Rabinowitch, R. Revelle, E. Staley, E. P. Wigner, J. R. Zacharias **UK**: E. Bullard, B. Lockspeiser, J. Rotblat **AUS**: J. Crawford **BUL**: G. Nadjakov **Brazil**: C. Pavan **Italy**: G. Bernardini **NE**: B. V. A. Röling

[14] *Proceedings of the 7th Pugwash Conference, Stowe, September 1961*. 5–9 September 1961. At both Stowe conferences, Interpreters (Russian/English) were assigned to each Group—from a five-strong pool comprising N. Habarin, F. Sviridov, V. Pojidaeff, N. Bevad and L. Maksimov.

in Europe near CERN, in or in Denmark or Austria, to better compete with the US in the particle physics field.[15] Members of this group included Nikolai Bogoliubov, Mikhail Dubinin and Igor Tamm from the USSR, the Americans Jay Orear, Isador Rabi, Alvin Weinberg and Jerrold R. Zacharias, the CERN-based Italian Gilbardo Bernardini, the Japanese physicist Toshiyuki Toyoda and Joseph Rotblat. Zacharias complained that focusing on high energy physics showed a "lack of imagination"—as he put it, "CERN okay, so do it again"—to which Rabi and Tamm responded by suggesting large projects centered on computers and semi-conductors respectively. Elsewhere, Tamm and the West German Gerd Burkhardt floated an idea for collaborative research in plasma physics. All agreed on any such project being 'intercontinental'—meaning, having global reach—with one suggestion for an Indian location, an acknowledgement of the need to support science and technology within the "developing" countries of the Global South. Cooperation with this region was handled specifically in Working Group 6. The statement issued from this conference emphasized such cooperation to be both a "duty and a necessity" with the caveat that the means and mechanism of cooperation be "rendered so that it would not impair the independence of any country." All of this was cast in the familiar Pugwash narrative that scientific cooperation "could help to reduce world tensions, to strengthen peace, and to further disarmament" (Rotblat 1967, 118–119).

Based on plenary discussions and the findings of the Working Groups, the conference identified some fifty projects "considered worthy of international collaboration" although many were not original and were already under consideration, elsewhere, for example, within the International Council of Scientific Unions (Rotblat 1967, 30, 118). The list ranged across the disciplinary spectrum, including the earth sciences, human health, space exploration, including "internationalizing the moon" and "grappling with human pollution and the problems of the environment." Two projects stood out. The first, tabled in a plenary session by Martin Kaplan (1915–2004) and Bernard Röling, was an idea for an Inter-Continental Science Center. This proposed a "cluster" of laboratories and institutes housing "Big Science" facilities, including particle accelerators and computers, in Berlin—which might have "extraordinarily great significance" in improving the political climate between East and West and which could be developed under the auspices of the United Nations.[16] The suggestion of Berlin as the location for this project elicited sharp criticisms from almost everyone as being naïve and utopian—all recognized the impossibility of this until the German question was resolved—even if Rabinowitch and the West German Gerd Burkhardt remained enthusiastic.[17] In the event, although the idea for an intercontinental science center in Berlin was raised in Bonn by Werner Kliefoth and continued to run within Pugwash for several years, the idea came to nothing (Kraft 2020). As Rotblat reported to the Continuing Committee in 1963, his discussions with scientists

---

[15] Minutes from meeting of Working Group 5 on "physical sciences," 6 September 1961. RTBT 5/2/1/7 (13).

[16] *Proceedings of the 7th Pugwash Conference, Stowe, September 1961:* 185. 'International cooperation in science.' Notes on 4th Plenary session, afternoon of 8 September 1961. RTBT 5/2/1/7 (13).

[17] 'International cooperation in science.' Notes on 4th Plenary session, afternoon of 8 September 1961. RTBT 5/2/1/7 (13).

from East and West Germany had "shown that the present time is not opportune for calling a conference in Germany on the setting up of a scientific center in Berlin."[18]

Martin Kaplan was also behind a second noteworthy idea, this time arising out of Working Group 4, which dealt with international cooperation in the life sciences. Discussed under the header "Endless frontiers"—a play on the metaphor made famous in 1945 by Vannevar Bush—this group envisaged an "explosive" growth in the biological sciences in the coming decades, especially in the areas of genetics and/or molecular biology. Comparisons were drawn with the transformation of physics early in the century. Anticipating medical advances, Kaplan argued a corresponding need for large cooperative projects in the health sciences, perhaps administered by the World Health Organization (WHO), the "major operating international agency in the medical field" (Rotblat 1967, 15). Kaplan's ambitious plan envisaged multiple institutes and cutting-edge laboratories carrying out a wide-ranging research program into a wide spectrum of diseases, including those afflicting the "developing world," involving many medical specialisms and with facilities for training and conferences. This medical "mecca" would operate under the umbrella of the WHO—where Kaplan was at this time based—but with sponsorship from organizations such as the UN. This proposal was taken up in 1963 by WHO director Marcolina Candau, but the project became mired in long-running debate within international science policy circles (Rose 1965; Calder 1965; McElheny 1965). In the end, it floundered on the central difficulty that bedeviled all international science projects, namely, the trade-offs/transaction costs of cooperation, and the need to satisfy the national interests of participating countries (Rose 1965, 349–351; Calder 1965). In this case, the British were against the project and in frustration, in May 1965 Rotblat wrote to *The Times* in support of it. In the event the plan was finally abandoned at the eighteenth World Health Assembly meeting in Geneva later that month (Leach 1965).[19]

Indeed, most of what was a utopian list of possible international projects discussed in Stowe remained on paper. Nevertheless, at the time, these ideas were collected together in a report Rabinowitch described as "a kind of charter for international cooperation." From time to time, the "Stowe Report" was a subject of discussion within the PCSWA as promoting international scientific cooperation became in the 1960s an established part of the Pugwash agenda (Rabinowitch 1965, 11). Externally, the PCSWA's promotion of international cooperation was favorably received. For example, in a letter to Martin Kaplan soon after Stowe, Joseph Rotblat reported that "Rabi was extremely impressed with the first week and he rang up to say that the people high up in Washington thought very highly of our statement."[20]

The first Stowe conference was significant for both Rabinowitch and Kaplan. For Rabinowitch, it furthered his vision for Pugwash to promote international scientific cooperation. Henceforth, this theme was included on conference programs and was

[18] Minutes of 16th Pugwash Continuing Committee, 8–10 February 1963, Ciba, London, 1. RTBT 5/3/1/2 (Part I) (4).

[19] Joseph Rotblat. Letter to *The Times,* 10 May 1965. RTBT 5/2/1/4 (Part II of II).

[20] Joseph Rotblat to Martin Kaplan, 22 September 1961. RTBT 5/2/3/5.

addressed within Working Groups. Beyond foregrounding the distinctive expertise offered by the PCSWA, as Bernard Feld emphasized, international scientific cooperation was "not adequately" covered by any other organization.[21] Feld and Rabinowitch saw this as one means to strengthen engagement with the countries of the so-called developing world. In practice, however, the Pugwash organization struggled both to build links with the Global South and to increase the participation of scientists from this and other "developing" regions. The apparent successes of conferences held in Udaipur and in Addis Ababa in 1965 and 1966 proved difficult to build on. Moreover, some colleagues were less convinced that the organization could or should move in this direction. Advances here, including the formation of a Pugwash study group on development, had to wait until the early 1970s.[22]

Martin Kaplan's profile within Pugwash was strengthened at Stowe. A veterinarian and microbiologist, he first learned about the organization from his Scottish friend, the journalist and peace activist Ritchie Calder. In 1958, Kaplan attended the third conference in Vienna together with Brock Chisholm, the director of the WHO, where Kaplan was based. Close to both Rotblat and Feld, Kaplan had a wealth of experience and expertise in the realm of international scientific cooperation and was well-connected within science policy networks. Working with Rotblat and James Wise, he had been heavily involved in the creation in 1959 of the European Pugwash Group, and took an active role in the group, sometimes hosting its meetings in his Geneva home. That year he was also instrumental in organizing the fifth Pugwash conference dedicated to Biological and Chemical Weapons, which was a landmark for international discussions on this topic. He went on to play a decisive role in the creation in 1964 of the Pugwash Study Group on these weapons. Although for many years Kaplan held no official post within the organization, he steadily became a part of its senior 'inner circle.' Much later, between 1976 and 1988, he served as Secretary General.

Shortly after Stowe in October 1961, a full report of both conferences was published in *Science,* the official journal of the *American Association for the Advancement of Science*, preceded by a summary overview by Glass (1961). In reviewing the many ideas for international cooperation, Glass sounded a note of caution, conceding that, "...in the fine glow of such hopes, every mind harbored the unspoken recognition that none of these joint efforts could germinate in a world on the eve of nuclear war" (Glass 1961, 984). The sense of that world was strongly apparent in the conference the following week. Although many of the same scientists were present, their attitude, and the atmosphere around the table, were very different.

---

[21] Bernard Feld to Joseph Rotblat, June 1966. RTBT 5/2/1/16 (33).

[22] Minutes of the 33rd meeting of the Continuing Committee, 10–16 September 1970, Fontana, US. RTBT 5/3/1/2 (7).

3.3   The Second Stowe Conference: Acrimony and Cold War Grandstanding

The second conference in Stowe proved an altogether more difficult affair, given that its theme, 'Disarmament and world security' took those present into politically incendiary terrain, not least the weapons testing controversy. Immediately, the tone was set in the official messages of support from President Kennedy and Premier Khrushchev. Both mentioned the Soviet resumption of weapons tests, Kennedy speaking of the "somber turn of events" in the past week (Kennedy 1961, 7). On the offensive, Khrushchev explained that this was one of a "number of measures to strengthen the security of our country, to block the way for German revanchists and their patrons," and that this step had been taken "with a heavy heart and deep regret" (Khrushchev 1961, 9–10). Of course, he added too that "the peaceful policy of the Soviet Union remains unchanged."[23]

In his opening remarks, Rotblat emphasized what he saw as the success of the past week, before moving swiftly to acknowledge the challenges ahead, warning that it would be "foolish to expect too much."[24] But—ever the optimist—he was hopeful that "the least we might achieve is that each side will get a better understanding of how the other side views the crisis." Bringing his own distinctive brand of conciliatory diplomacy to bear on the situation, he called for the conference to abide by the rules governing discussions at Pugwash, namely: "to tell the truth, without being abusive; to be frank, but not to embarrass" and, overall, to retain the "strange mixture of frankness and friendliness, restraint and directness, that Pugwash has somehow hitherto managed to produce—which some have called the 'Pugwash spirit'."[25] In the week that followed, this "spirit" was in very short supply.

By far the largest delegation—twenty-three strong—was that of the Americans, with eleven scientists each from the Soviet Union and from the UK (Rotblat 1967, 125–126).[26] The Soviet delegation remained unchanged from the previous week, but there were changes in the American party: incoming were Donald Brennan, Amron Katz, George Kistiakowsky and Henry Kissinger. Beyond the by now usual stalwarts, the British delegation included Whitehall and establishment figures, notably Solly Zuckerman, Sir Edward Bullard, Michael Howard—Alistair Buchan's colleague at the ISS—and, for the first time, two of the country's top nuclear scientists, John Cockcroft and William Penney. As in the previous week, Australia, Austria, Belgium,

---

[23] The official message from the Soviet Peace Committee reiterated Khrushchev's points, especially regarding West Germany. Message from the Soviet Peace Committee, 11.9.1961. *Proceedings of the 8th Pugwash Conference,* Stowe, September 1961: 11.

[24] Joseph Rotblat, Opening remarks at the eighth conference, Stowe, VT. Notes on the 1st Plenary Session, 11 September, 1961, 9.30 am–12.30 pm. RTBT 5/2/1/8 (1).

[25] Rotblat, Opening remarks at the eighth PCSWA, Stowe, VT, 11 September 1961. RTBT 5/2/1/8 (1).

[26] These were the 'total' number of participants per country, and included those categorized as 'observers,' which numbered four, three and four for the US, USSR and UK respectively. The Soviet observers at both Stowe conferences were: N. I. Bazanov, S. G. T. Korneev and Vladimir P. Pavlichenko.

Canada, France, Hungary, Japan, the Netherlands and West Germany were represented by one or two scientists. To the disappointment of all, the Chinese were absent and amid the worsening relations between China and the Soviet Union, concern deepened that the Chinese were disengaging from the PCSWA (Barrett 2020; Rotblat 1967, 109–127).

As in Moscow, the plenary program was dominated by the Soviets and Americans; the opening panel featured papers by Khvostov, Doty and Topchiev.[27] Solly Zuckerman's assessment of these three key players is illuminating: Khvostov he found "aggressive," whilst Doty was "quiet and reasonable" and he found Igor Tamm generally more conciliatory, even pleasant; of Topchiev, he said only that his paper was a justification of the Soviet resumption of tests.[28] Zuckerman's British colleagues, Cockcroft and Penney, gave a joint paper.[29] In applying their scientific and technological expertise to the intractable problems of disarmament, these scientists were sharing their technical know-how, but always within limits: if the emphasis was on science, the wider context of what was being discussed was inherently political. Delegates could also take advantage of a disarmament library comprising relevant background papers by, for example, Jerome Wiesner (who was not in Stowe) and Patrick Blackett, various disarmament-related publications of the US Department of State, and recent books about disarmament by, for example, Donald Brennan, Thomas Schelling, and Louis B. Sohn.[30]

Zuckerman was right about Topchiev's paper. The collegiality of a week earlier was gone, replaced by a Cold Warrior rhetoric in tune with the Kremlin line. Echoing Khrushchev, Topchiev emphasized that with a "heavy heart and deep regret" the Soviet Union had been "compelled" to resume tests by the changed security situation. Emphasizing the transformation of Germany into what he called a militaristic state "armed to the teeth" by Chancellor Adenauer and his "patrons," Topchiev explained that "We are openly threatened by force and we have nothing left but to answer with force." He concluded by calling for the German question to be settled and for a German peace treaty (Topchiev 1961b). Bernd Kubbig reports that Topchiev was "in regular telephone contact with Moscow," and also that "he kept a close watch on the smaller groups when the plenary sessions broke up" (Kubbig 1996). If Topchiev was regarded as a crucial interlocutor by his western colleagues on the Continuing

---

[27] Vladimir Khvostov (Review of disarmament since Moscow); Aleksandr Topchiev (General and Complete Disarmament); Paul Doty (Disarmament) and Harrison Brown/Amron Katz (An approach to disarmament).

[28] Solly Zuckerman. Report on Pugwash conference in Stowe. Ministry of Defence, 26 September 1961. RTBT 5/1/3/7. [Marked secret/UK eyes only. Date stamped at John Cockcroft's office 29 September 1961]. I am indebted to Lawrence Wittner for depositing this source in the Rotblat collection, and for his kind permission to use it.

[29] John Cockcroft and William Penney. Cessation of the production of fissile material/Elimination of weapons stockpiles. Program details in: COSWA Proceedings VIII, Draft, 22 February 1962. RTBT 5/2/1/7 (9).

[30] The titles of the Wiesner and Blackett papers were 'Inspection for disarmament' and 'Critique of come contemporary defense thinking,' respectively.

Committee, for him his role at Pugwash was always a careful balancing act vis-à-vis his relations with the Kremlin and any loyalty he had to the project (Lüscher 2020).

But, in a disclosure that casts light on why Topchiev was seen as important by his Western colleagues on the Continuing Committee, he went on to reveal details of private exchanges with American colleagues ten months earlier in Moscow. Here, as he put it, the Americans had emphasized "the need for a transition period to the Kennedy administration" which, once in power, would develop policies to reduce tensions and that, in light of this, for the time being, it would help if the Soviet Union could refrain from anything that could cause difficulties (Topchiev 1961b; Kubbig 1996, 9). Topchiev's candid revelations about off-the-record conversations cast light on how scientists from East and West were able privately within Pugwash to create between and around them a space where a degree of autonomy was possible. That such understanding was possible points to the trust-building process made possible by Pugwash and which, at the same time, constituted a novel form of informal diplomacy between scientists. On the one hand, this rested on tacit understanding both of each other and of what together they were trying to achieve. On the other hand, the possibility for such moments of mutual understanding was contingent in part on the wider political context, namely the state of play between the superpowers. In September 1961, acute tensions in this relationship made the second Stowe conference an occasion for Cold War grandstanding, including by the respected Topchiev.

Nevertheless, the Working Groups went ahead. These are summarized in Table 3.2 which affords a sense both of the topics discussed and their cross-bloc character (Table 3.2). It makes clear that western scientific advisors close to Washington and Whitehall such as Paul Doty and Solly Zuckerman were actively involved in small group work where they mixed and mingled with their Soviet counterparts in private and out of the spotlight. As became the convention, each group produced a report which was circulated and discussed in plenary session towards the end of the conference—creating a resource of use within and beyond Pugwash. Working Group 1 proceeded smoothly and its report was completed promptly; its members then joined other groups. Groups 2 and 4 reached agreement in broad terms—but without any major breakthroughs. By contrast, Group 5 failed to reach agreement on anything: in Wayland Young's view, this "ran head on into an ideological conflict and never got off the ground."[31] Working Groups could provide a useful barometer as to what was and what was not possible at a particular moment in East–West relations. Overall, the positive outcomes from some groups enabled Glass to claim in his report on the Stowe conferences in *Science* a degree of success for the eighth conference even in a period of deep political crisis (Glass 1961).[32]

---

[31] Report of Working Group 5, *Proceedings of the 8th Pugwash Conference, Stowe, September 1961*: 29–32, especially 29. Wayland Young, Report on the Stowe Conferences, 1961. RTBT 5/2/1/8 (1).
[32] Reports of Working Groups 1–5 at the eighth PCSWA, Stowe, VT, US. 11–16 September 1961. RTBT 5/2/1/8 (1).

At any rate, it was immediately clear that Working Groups added a valuable dimension to Pugwash, creating another private space for informal discussions. In his internal report on Stowe, Wayland Young reported that "the real work of the conference went on in the Working Groups"—a perceptive observation that anticipated the value of these fora for the future work and reputation of Pugwash as a forum for East–West dialogue.[34]

Table 3.2 Working Groups: Eighth Pugwash Conference Stowe, VT, USA, 11–16 September 1961[33]

Group	Topic	Members
1	Nuclear production and stockpiles	**UK**: J. Cockcroft, P. Noel-Baker W. Penney **US**: H. Bethe **USSR**: M. M. Dubinin, N. I. Sissakian, I. E. Tamm **FRA**: F. Perrin **Rapporteur (R)**: E. Young
2	Delivery systems	**UK**: E. Bullard **US**: R. R. Bowie, D. Brennan, T. Gardner, G. Kistiakowsky **USSR**: A. A. Blagonravov, N. N. Bogoliubov, S. G. T. Korneev **R**: Amron Katz
3	Selected next steps in arms reduction	**UK**: P. M. S. Blackett **US**: P. Doty, B. T. Feld, W. K. H. Panofsky **USSR**: V. P. Pavlichenko, N. A. Talensky **Canada**: J. Polanyi **FRG**: G. Burkhardt **Japan**: T. Toyoda **R**: Ruth Adams
4	Problems of GCD	**UK**: M. Howard **US**: H. Brown, B. Glass, E. M. Purcell, L. B. Sohn **USSR**: V. M. Khvostov, A. V. Topchiev **Australia**: M. Oliphant **FRA**: P. Rosenstiehl **NE**: B. V. A. Röling **R**: Patricia Lindop
5	Preconditions to possible success at negotiations and initiation of disarmament measures	**UK**: J. Rotblat **US**: C. Lauritsen, L. Lipson, I. I. Rabi, E. Rabinowitch, C. Townes **USSR**: N. I. Bazanov **AUT**: H. Thirring **HUN**: F. B. Straub **R**: M. Sands

---

[33] *Proceedings of the 8th Pugwash Conference, Stowe, September 1961.* 11–16 September 1961.

[34] Wayland Young, Report on the Stowe Conferences, 1961. RTBT 5/2/1/8 (1).

Young's report on this conference noted Topchiev's staunch defence of the resumption of tests and recounted tense exchanges about this. He also recounted discussions amongst western participants about whether the Soviet scientists had known in advance about this—an issue which remained unclear.[35] Elsewhere, US-Soviet antipathies in Stowe were noted in a report filed by the West German physicist Gerd Burkhardt with the Vereinigung Deutscher Wissenschaftler (VDW).[36] Perceptively, Burkhardt argued that for all the tensions even the most sceptical participant would conclude that the second Stowe conference had been really "useful." Burkhardt captured an important point: the arguably unique value of Pugwash lay partly in its role as a site of disagreement and confrontation, and as a place for airing grievances.

Publicly available Pugwash sources do not cast much light on the tensions during the second week in Stowe. The short volume of the *Proceedings* contained papers from the plenary program and the reports of the Working Groups were not included.[37] Discussions about the conference statement ended without unanimous agreement and, for the first time, involved abstentions—on the part of five Americans.[38] In his internal report on Stowe, Wayland Young reported fierce arguments about the content of the statement, with the Soviets adamant that there should be no mention of the test issue, something which the Americans would not countenance. As a compromise, a short communique style statement was produced, essentially limited to an outline of the program. As Young noted this meant that, in the end, the statement did not "express much more than a common hope for peace." Resorting to expressing banalities of this kind ran the danger of making Pugwash look weak and of undermining its role and reputation.

The posturing along bloc lines seen in Stowe showed that for all the rhetoric to the contrary, the Pugwash organization never could always or fully overcome the divides it sought to transcend. The myth that scientists could suspend political and national allegiances was part of the rationale for Pugwash and a useful narrative in building its identity and role. But as Eugene Rabinowitch acknowledged, "like all citizens of nation states" scientists "are swayed by national beliefs and animosities." Being 'on the ropes' in Stowe was part of the process by which Pugwash scientists were 'learning the ropes' of the difficult international role they aspired to in an organization that sought to bridge the divide between East and West.

In his report of both conferences in *Science* in October 1961, Bentley Glass alluded to the difficulties during the second week in Stowe, conceding that it had not been easy for participants "to speak without mutual recrimination or anger." In his view

---

[35] Wayland Young, Report on the Stowe Conferences, 1961. RTBT 5/2/1/8 (1).

[36] Gerd Burkhardt, Report on the eighth PCSWA in Stowe, VT, USA. Copy in File B43.12, Auswärtiges Amt, Berlin. In April 1962, Burkhardt sent his report to Dr. Gerhard Schröder, Bundesminister für Auswärtige Angelegenheiten.

[37] Minutes of the thirteenth meeting of the Continuing Committee, 2–15 September 1961, Stowe, VT, USA. RTBT 5/3/1/2 (Pt 1) (3). In Moscow and for the eighth conference in Stowe, two versions of the *Proceedings* were produced, with and without the minutes of plenary discussions. The Continuing Committee presided over which versions were sent to whom. For example, in terms of the eighth conference, heads of state received the copy without the discussions.

[38] The Americans abstaining were: Robert R. Bowie, Donald Brennan, Henry Kissinger, Amron Katz and Leon Lipson. Absent during the discussion of the statement were John Cockcroft, Trevor Gardner, Charles Lauritsen and Isador Rabi.

rendered, this it "all the more significant that the participants were able to arrive at constructive recommendations. As befits scientists, views from East and West were exchanged with frankness and reasonable objectivity" (Glass 1961). For this public audience, Glass was glossing the extent of the disagreements and the intensity of the arguments around the table in Stowe. Perceptions mattered, and it was important to portray Pugwash and its conferences as a stable, discreet and reliable forum of East–West communication.

But there were some positives to take away from Stowe. The report of Working Group 1 attracted interest in government quarters in the west. In March 1962, Zuck-erman wrote to Rotblat to ask whether the British Foreign Office could table part of it at the upcoming session in the disarmament talks in Geneva. Rotblat granted permission and Zuckerman said he would pass the report on through "formal chan-nels" to the Americans for their consideration at Geneva.[39] This episode points to an interplay between unofficial and official diplomacy whereby those involved in offi-cial disarmament talks in Geneva could and did draw on the work of Pugwash and its scientists—a development that likely bolstered the reputation of the organization as a serious and credible actor on the international stage of nuclear and disarmament diplomacy.

A further indication that potentially useful conversations had happened in Stowe was apparent in a letter from Patrick Blackett to Joseph Rotblat in late November 1961 discussing plans for the upcoming PCSWA in Britain.[40] Making a point about what he saw as the difficulty of having meaningful discussions about disarmament between experts "in a big conference under the glare of publicity," Blackett reminded Rotblat of an episode in Stowe when "Kissinger strongly urged no precipitate publica-tion" of some of the discussions quoting Kissinger as having said, "In fact, if you can keep quiet" (about these discussions) "we can try and sell them to the White House before a counter attack develops from the Pentagon building." Blackett's recollec-tions suggest the high-level relevance of the work on disarmament in Stowe and that this was reaching deep into senior political and policy-making circles in Washington. But they point also to how conversations and ideas went unreported in the internal records of the Pugwash organization. These omissions and silences protected its posi-tion, and those of its scientists, within an informal circuitry of diplomacy operating between Pugwash scientists, and between the PCSWA and those making decisions in government.

Zuckerman's request for the Working Group report and Blackett's revelations about Kissinger's remarks in Moscow suggest that the Pugwash organization was taken seriously within senior government circles in London and Washington. Recog-nition was growing that its conferences could be active sites of exploratory, non-committal diplomacy that could move beyond what was possible within official channels. This different form of unofficial dialogue could complement and/or further what was taking place in formal channels of official diplomacy.

[39] Solly Zuckerman to Joseph Rotblat, 1 March 1962. Rotblat to Zuckerman, 2 March 1962. Zuckerman to Rotblat, 5 March 1962. RTBT 5/2/1/7(9).

[40] P. M. S. Blackett to Joseph Rotblat, 29 November 1961. RTBT 5/2/1/10 (28).

But in Stowe, Pugwash had stood on the brink of failure. In a letter to Martin Kaplan shortly afterwards, Rotblat confided that the acrimony in Vermont had been such that, "it looked as though Pugwash was finished."[41] He went on to concede that the "hopeful prognosis" from the hard work at Moscow "appears to have been dashed." Instead of capitalizing on the advances in Moscow, Pugwash had experienced a stark lesson in how an outbreak of East–West hostilities could destabilize the organization. But Rotblat's unwavering commitment to the project remained intact and he remained hopeful, confiding to Kaplan that if Pugwash could survive "this episode," then "I think we are going to stay."[42] On this account, the eighth conference had stretched the goodwill between its scientists to the limit and in hindsight had provided a major test of the resilience of the PCSWA.

Rotblat was in part echoing Gerd Burkhardt's point about the usefulness of arguments across the Pugwash table. In the polarized Cold War setting, disagreement was part and parcel of the process of building relationships between Pugwash scientists. Surviving open enmities and discord was a part of the trust-building process. At one level, the Pugwash project was only ever held together by the strength of loose ties between its scientists from East and West—ties which, Stowe had shown, had been able to withstand anger and disagreement.

3.4 The Zuckerman Intervention: The British Change Tack on the PCSWA

On a more upbeat note, and in an important development for Pugwash in the UK, Rotblat reported to Kaplan that Cockcroft and William Penney had been impressed at Stowe and "felt strongly that it should be kept going."[43] Both Cockcroft and Zuckerman subsequently lent their support to a plan for the next conference in 1962 to be held in the UK—with Cockcroft even taking a place on the organizing committee.[44] For Zuckerman and Cockcroft, the eighth conference had provided a first opportunity to observe Pugwash and its scientists in action and in particular to witness first-hand the encounter between the Americans and Soviets during a conference riven with tensions and hostility. Perhaps being privy to arguments and confrontation in Stowe, but also to the connection and relationships between those arguing around the table, had persuaded them that something potentially useful was going on here.

Certainly, such views were apparent in Solly Zuckerman's report on Stowe for the Foreign Office, a copy of which he sent to John Cockcroft.[45] This report—classified at the time—is of particular interest because his assessment fed into changing

[41] Joseph Rotblat to Martin Kaplan, 22 September 1961. RTBT 5/2/3/5.

[42] Rotblat to Kaplan, 22 September 1961. RTBT 5/2/3/5.

[43] Joseph Rotblat to Martin Kaplan, 22 September 1961. RTBT 5/2/3/5.

[44] Minutes of the fourteenth meeting of the Continuing Committee, London, 26–28 January 1962. RTBT 5/3/1/2 (Pt 1) (3).

[45] Solly Zuckerman. Report on Pugwash conference in Stowe. Ministry of Defence, 26 September 1961. RTBT 5/1/3/7. [Marked secret/UK eyes only. Date stamped at John Cockcroft's office 29 September 1961].

views of the Pugwash organization and its conferences in British government circles. Acknowledging the opportunities for a "valuable exchange between ourselves and the Russians," Zuckerman spoke of a "transformation" such that, in his view, the PCSWA was now a "serious, if informal" site for "sub-governmental diplomacy."[46] This was a remarkable turnaround in Zuckerman's position. Only the year before, he had been ambivalent, even suspicious of the organization. Seemingly, he was now less concerned with the dissenting tradition of the PCSWA and more interested in it as a potentially valuable asset to the British government in the realm of nuclear diplomacy.

Zuckerman's views carried weight in Whitehall. Six months after Stowe, in April 1962, Duncan Wilson at the UK Foreign Office wrote a memorandum emphasizing the usefulness of Pugwash conferences which, as he put it,

> afford an opportunity to smoke senior Soviet disarmament experts out, on an informal basis, about the official position their government is taking in the formal disarmament negotiations. It is not inconceivable that this could lead to some break-through. At all events, we are convinced the chance should not go by default.[47]

The Pugwash organization and its conferences were registering in new ways in Whitehall. This changed attitude was apparent in a report from within the Ministry of Defence in 1963 which emphasized that its "real value" lay "in providing a window on the communist world—not the one-way window of which we have heard so much, but a two-way window through which we can both observe the communists, and allow them to sample life in the free world."[48] Having once dismissed Pugwash as a tool of Soviet propaganda, the British Foreign Office saw no irony in seeking now to use it for Western propaganda. More than this, they now perceived the organization as affording rare opportunities for informal diplomacy and from this, the possibility that it could be useful to the interests of the British government.

Zuckerman also noted the "quasi-official" status of a large part of both the American and Soviet delegations. Indeed, he cited one American who had made the observation that, "Before, if one went to these meetings, one lost all one's "clearances;" to come to this one, one practically needed a special pass to show that you were a trusted man."[49] This resonated with the views of Eugene Rabinowitch who also discerned qualitative changes in the PCSWA after Moscow and Stowe which, as he put it, were taking on the character of "semi-official get-togethers of scientific experts which are trying to find in private discussion a way out of the deadlocked

---

[46] The eighth conference in Stowe had been Zuckerman's first: but through his role at the Foreign Office he was likely privy to assessments of and discussions about the Pugwash organization—which possibly provided the basis for his assessment that it had undergone a "transformation."

[47] A. D. Wilson to C. J. Hayes, Treasury. 26 April 1962. FO 371/163160. RTBT 5/1/3/7. (This document is amongst the material deposited at the Churchill Archive Center by Lawrence Wittner.)

[48] B. T. Price. Report on Dubrovnik Conference, 09/1963. 16 October 1963. Marked confidential. FO 371/171/90. In: RTBT 5/1/3/7.

[49] Solly Zuckerman. Report on Pugwash conference in Stowe. Ministry of Defence, 26 September 1961. RTBT 5/1/3/7. [Marked secret/UK eyes only. Date stamped at John Cockcroft's office 29 September 1961].

disarmament negotiations." He went on to note that "The important positions some of the American participants hold in Washington has inevitably affected the style of the discussions" (Rabinowitch 1963). To illustrate his point, Rabinowitch recalled how one of the new Russian participants was recently moved to remark that he had thought the Pugwash conferences to be "officially unofficial," but that they now (in 1962) appeared to him "unofficially official" (Rabinowitch 1963).[50]

Rabinowitch was perhaps not altogether at ease with this shift. Gaining a reputation as being useful to state actors, and moving closer to government through its emerging role on the international stage as a forum for disarmament conversations, had far-reaching implications for Pugwash. Not least, it raised questions about how the organization and its scientists would maintain a critical edge vis-à-vis state actors. Of course, the PCSWA had constantly to develop and change. But Rabinowitch seemed concerned that moves in this direction had to be carefully managed so as to protect the organization's much-vaunted and cherished independence and autonomy. In effect, this difficulty flowed from the paradoxical vision of Pugwash, namely, its strategy of working both with and against governments—which, as time passed, grew more complex.

The eighth conference, especially when contrasted with Moscow, underlined to senior Pugwash scientists that the learning curve inherent in the East–West project they were embarked upon was both steep and unpredictable—and that its gradient was largely set by geopolitical developments beyond their control. In navigating the rollercoaster of the Moscow and Stowe conferences, new patterns were emerging in the internal dynamics of the PCSWA. Joseph Rotblat played a key role in steadying the organization—amplifying further his influence within it. As Secretary General, he enjoyed executive and discretionary powers, a permanent seat on the Continuing Committee and a good deal of autonomy. He intuitively understood that remaining neutral and, moreover, performing impartiality were key to establishing and maintaining the power, authority and respect accorded this office within the organization. An impartial Secretary General, even-handed in respect to East and West, was vital to how the PCSWA functioned internally and helped Pugwash to work as a cross bloc project.

One can, for example, point to the pattern of Rotblat's participation in Moscow and Stowe.[51] Beyond his opening addresses as Secretary General, Rotblat remained largely silent. In what would become characteristic for him at conferences, including within working groups, the available reports and accounts indicate that he remained on the periphery of discussions, adopting instead a role as neutral observer. This did not mean that he was in any way passive. On the contrary, this was part of a process through which he was actively developing his own brand of citizen diplomacy within the organization. Adept at working discreetly and informally 'behind

---

[50] This comment was made to Rabinowitch at the ninth PCSWA held in Cambridge in the UK in September 1962.

[51] A glance through the *Proceedings* for the Moscow and Stowe conferences reveal Rotblat's absence in discussions at sessions and in working group meetings at which he was present. Primary sources, including the minutes from the Stowe conferences, also echo this pattern.

the scenes,' Rotblat was honing his skills as effective intermediary within and across the blocs, liaising, brokering, and generally orchestrating a great deal of the business carried out at and between the conferences. Very little went on that he did not know about. All of this meant that, in practice, Rotblat stood somewhat apart from many of his colleagues—although this did not preclude close relations with some notably, but not exclusively, Americans including Feld, Kaplan and Rabinowitch. Looking Eastwards, Rotblat believed the Soviets believed in the Pugwash project and there was great respect and goodwill between Rotblat and Topchiev. In a sense, Rotblat embodied the founding East–West spirit and principles of Pugwash. This was important as Rotblat took charge of the plans for the two conferences in Britain that followed Stowe.

In Stowe, the Pugwash leadership expanded the in-house repertoire of communication channels at conferences with the introduction of Working Groups. This innovation had transformative effects on the organization, in terms both of the quantity and quality of work accomplished during conferences, and in fostering personal links and networks been Pugwash scientists. The seeds for these groups had been sown in Moscow, where small break-out groups had been a pragmatic response to the need for additional opportunities for in depth, informal discussion. In Vermont, the protocols governing these groups rested on Chatham House rules which enabled scientists from East and West to work closely together in an informal and private setting. Groups were also required to produce reports on their work, which became a valuable resource within and sometimes beyond Pugwash. Henceforth, Working Groups constituted a major locus of the work carried out at the conferences. Importantly, they also provided another context for the kinds of encounters and transnational exchanges that fell under the rubric of informal back-channel diplomacy. These groups also lent a more decentralized structure to the PCSWA, functioning as internal 'hubs' of activity. On occasion, they could serve as forces for change within the organization, for example, affording Pugwash scientists a means to press for particular topics to be added to its agenda. Sometimes too these groups could challenge the authority of the Continuing Committee.

The Pugwash organization and its conferences were a prism through which the state of the superpower relationship—and the wider Cold War—at any one time were refracted. Each conference had its own dynamics, as the high of Moscow and the low of Stowe had shown. Over the course of eight very different conferences, the leadership had learned a great deal about operating at the intersection between science and politics, equipping its senior figures with a unique fund of experience and practical know-how about the 'rules of engagement' when tackling the problems of nuclear disarmament. These 'rules of engagement' shaped the *modus vivendi* between American and Soviet scientists of Pugwash as they learned how to work across the Cold War divide. In the course of these encounters and transnational exchanges they were developing an innovative form of technopolitical communication which harnessed the tools and conventions of scientific communication to the analysis of international political problems relating to disarmament. In turn, this was the basis for a still-emerging, distinctive brand or style of informal diplomacy. These were resources on which the leadership drew as the organization sought to move beyond Stowe.

3.5   1962: Cambridge and London

It had been confirmed at the meetings of the Continuing Committee held during the Stowe conferences that the next conference would take place in 1962 in London, and that this would be a large conference to mark the fifth anniversary of the PCSWA.[52] The idea was that this would be an opportunity to review the organization and plan the next five years. The British also pressed for a second smaller conference in the UK, to precede that in London, which would be focused on disarmament so as to continue the conversations begun in Moscow. Still unnerved by the difficulties and loss of momentum in Stowe, Harrison Brown suggested that the British conference be delayed for a year. This was rejected outright by an uncharacteristically angry Rotblat, on the grounds that plans for both conferences were well-advanced, and that with John Cockcroft on the UK planning committee, a delay was out of the question.[53] The ninth and tenth conferences took place in August and September 1962, in Cambridge and London, respectively.

The ninth conference, focused on 'Problems of disarmament and world security,' took place in Caius College at the University of Cambridge, and was attended by sixty-five scientists, including many who had been in Stowe and/or Moscow.[54] Most of the 'core' Soviet group of seven was there, and with the addition of Alexander M. Kuzin, J. Riznichenko and Igor Tamm. The American party included Brennan, Doty, Feld, Amron Katz, Leghorn, Sohn, and Kissinger. All four working groups were dedicated to disarmament topics. Henry Kissinger chaired Group 2 which was focused on 'Problems of balanced reduction and elimination of conventional arms'— the cross-bloc membership of which included Brennan, Bogoliubov, and Talensky. As hosts, the British were there in force, including Alistair Buchan, Edward Bullard and Alexander Haddow.

Opening the plenary sessions with a paper entitled 'The way ahead,' the British physicist P. M. S. Blackett struck an upbeat tone, framing the conference as an opportunity to build on the more positive outcomes from Stowe. In Blackett's view, despite the tense political crisis of that time, "some progress was achieved in the task of mutually understanding the detailed problems of disarmament, and the labours of the Working Groups led to some specific proposals for compromise on important questions" that had been accepted by scientists from both sides of the block divide (Blackett 1962, 39). Blackett hoped that, a year on and with the "international atmosphere" now "much calmer," progress could be made in Cambridge. To this end, he implored his colleagues in East and West to undertake a "mental role reversal" by

---

[52] Minutes of the thirteenth meeting of the Continuing Committee, 2–15 September, 1961, Stowe, VT. RTBT 5/3/1/2 (Pt 1) (3).

[53] Joseph Rotblat to Harrison Brown, 21 November 1961. RTBT 5/4/6/8. In a turnaround echoing that made by Solly Zuckerman, Cockcroft was now serving on the British Advisory Committee of the PCSWA—the predecessor of the British Pugwash group which was not formally established until December 1962—and which was the organizing body for the Cambridge and London conferences. RTBT 5/6/1/1.

[54] Bentley Glass. Report on the ninth conference. PCSWA 10, London, September 1962, Paper X.4.3. RTBT 5/2/1/10 (5).

visualizing "the armament and disarmament scene as it must appear from the other side." Blackett went on to detail the key developments in the disarmament realm since Stowe, especially the McCloy-Zorin agreement at the UN, which set out the principles for future disarmament negotiations. He concluded by arguing that the military policies of the great powers and disarmament policy were "two sides of the same coin" and should be studied together. Staying fully apprised of the 'state of play' in the disarmament conversation in Geneva and at the UN was for Pugwash scientists more than preparation for the conferences. Beyond their lives at the laboratory bench, this senior scientific elite followed events at this particular intersection between science and politics on a year-round basis—it was part of their lifeworld and worldview (Agar 2012).

In his report on the Cambridge conference, Bentley Glass briefly summarized the reports of the four working groups before concluding with an assessment of the state of play of disarmament dialogue within Pugwash, highlighting the increasing technical dimensions of this work:

> In conclusion may I say that if the advances in agreement seem small and the areas of difficulty seem large in these considerations of disarmament problems, it is in the first place because we are getting deeper into them and grappling with more technical matters than at first; and secondly, it is because we are advancing even farther beyond the matters considered by our official negotiators in Geneva. We have no reason for discouragement. Even the least advance in these considerations may make possible the breakthrough to a disarmed and peaceful world.[55]

The tenth Pugwash conference, held in London in September 1962 was planned and portrayed as a landmark anniversary, a celebration of the work of the PCSWA since 1957. With 175 participants from thirty-six countries, this was the largest Pugwash meeting to date. Henceforth all 'quinquennial' conferences—Ronneby in 1967, Oxford in 1972 and Munich in 1977—were accorded a special place in the Pugwash calendar, serving as an opportunity to review the past five years and to set priorities for the future (Rotblat 1972).[56] The quinquennial 'review and prospect' work in London was carried out by two Standing Committees, which considered 'future activities' and 'future organization' respectively.[57] The latter reaffirmed disarmament as the primary focus of the PCSWA, with three other themes—international scientific cooperation, scientists' social responsibility and the "developing" world—included on the agenda, but with lesser priority.

In his address as Secretary General, Joseph Rotblat shared his quiet but growing confidence in the Pugwash project, noting that the organization was beginning to garner "respect from the Establishment and from the scientific community," having

---

[55] Bentley Glass, 'Report on the ninth conference,' PCSWA 10, London, September 1962, Paper X.4.3, RTBT 5/2/1/10 (5).

[56] To this end, it became the convention for 'Standing' committees to be established in the run up to quinquennial conferences, completing 'retrospect and prospect' type reports that were pre-circulated and discussed during the conference.

[57] Reports from the Standing Committees, London, 1962. *Proceedings of the 10th Pugwash Conference, London, September 1962*, 'Proceedings of the eighth session,' 457–470.

acquired since its inception "goodwill, high reputation and vast experience."[58] Five years later at Ronneby, he would look back on London as marking the "peak of success" when, in his view, Pugwash had "proved itself" amongst scientists, politicians and the public and when, in his view, the earlier "indifference and suspicion" of state actors in the west had been replaced by "interest and sympathy."[59] In no small part this assessment was based on the sixth and eighth conferences held in Moscow and Stowe respectively.

But the second quinquennial period, 1962–1967, would bring new and serious challenges, as Pugwash struggled to respond to the changing political contours of the mid-1960s Cold War. Questions other than the nuclear arms race came to the fore. On the one hand, the Vietnam War made for increasingly uneasy relations between the PCSWA and the Lyndon B. Johnson administration in the US.[60] On the other hand, European scientists began to press for Pugwash to engage more strongly with the German problem. Both profoundly impacted the internal dynamics of the organization. Against this backdrop, members of the Continuing Committee also began in this period to mount informal back-channel initiatives beyond the conferences. In doing so, they took the Pugwash organization in new directions within the realm of informal diplomacy.

References

Agar, Jon. 2012. *Science in the twentieth century and beyond*. Cambridge: Polity Press.

Aronova, Elena, Karen S. Baker, and Naomi Oreskes. 2010. Big science and big data in biology: From the international geophysical year through the international biological program to the long term ecological research (LTER) network, 1957–present. *Historical Studies in the Natural Sciences* 40 (2): 183–224.

Barrett, Gordon. 2020. Minding the gap: Zhou Peiyuan, Dorothy Hodgkin, and the durability of Sino-Pugwash networks. In *Science, (anti-)communism and diplomacy: The Pugwash Conferences on Science and World Affairs in the early Cold War*, eds. Alison Kraft, and Carola Sachse, 190–217. Leiden: Brill.

Blackett, Patrick M. S. 1962. The way ahead. In *Proceedings of the 9th Pugwash Conference, Cambridge, UK, August 1962*, 39–54.

Bulkeley, Rip. 2000. The Sputniks and the IGY. In *Reconsidering Sputnik: Forty years since the Soviet Satellite*, ed. Roger D. Launius, John Logsdon, and Robert W. Smith, 125–160. Amsterdam: Harwood Academic Press.

Calder, Ritchie. 1965. The rejection of the WHO Research Center. *Minerva* 5 (4): 571–573.

Collis, Christy, and Klaus Dodds. 2008. Assault on the unknown: The historical and political geographies of the International Geophysical Year (1957–1958). *Journal of Historical Geography* 34 (4): 555–573.

Feld, Bernard A. 1973. A voice of conscience is stilled. *Bulletin of the Atomic Scientists* 4–12.

Fischer, David. 1997. *History of the IAEA: The first forty years*. Vienna: IAEA.

---

[58] Report of the Secretary General, *Proceedings of the 10th Pugwash Conference, London, September 1962*: 47–54.

[59] Rotblat at Ronneby. *Proceedings of the 17th Pugwash Conference, Ronneby, Sweden, September, 1967*: 21–28.

[60] Cecil F. Powell to Joseph Rotblat, 20 August 1966. RTBT 5/2/17/4 (Pt 3).

Glass, Bentley. 1961. Conferences on science and world affairs: The Stowe conferences on science and world affairs. *Science* 134 (3484): 984–991.

Haddow, Alexander. 1961. Cooperation in cancer research. In *Proceedings of the 7th Pugwash Conference, Stowe, September 1961*, 65–72.

Hermann, Armin, John Krige, Ulrike Mersits, and Dominique Pestre. 1990. *A history of CERN*. New York: Elsevier Science.

Kennedy, John F. 1961. Message to the eighth PCSWA. In *Proceedings of the 8th Pugwash Conference, Stowe, September 1961*, 7.

Khrushchev, Nikita. 1961. Official address to the conference, 5.9.1961. In *Proceedings of the 8th Pugwash Conference, Stowe, September 1961*, 9–10.

Kraft, Alison. 2020. Confronting the German problem: Pugwash in West and East Germany, 1957–1964. In *Science, (anti-)communism and diplomacy: The Pugwash Conferences on Science and World Affairs in the early Cold War*, eds. Alison Kraft, and Carola Sachse, 286–323. Leiden: Brill.

Kraft, Alison, and Carola Sachse. 2020. The Pugwash Conferences on Science and World Affairs: Vision, rhetoric, realities. In *Science, (anti-)communism and diplomacy: The Pugwash Conferences on Science and World Affairs in the early Cold War*, eds. Alison Kraft, and Carola Sachse, 1–39. Leiden: Brill.

Krige, John. 2016. *Sharing knowledge, shaping Europe*. Cambridge: MIT Press.

Krige, John, and Kai-Henrik Barth, eds. 2006. Introduction. Science, technology, and international affairs. *Osiris* 21 (1): 1–21.

Krige, John, and Luca Guzzetti. 1997. *History of European scientific and technological collaboration*. Luxembourg: OOPEC.

Krige, John, and Naomi Oreskes. 2014. *Science and technology in the global Cold War*. Cambridge: MIT Press.

Kubbig, Bernd W. 1996. *Communicators in the Cold War: The Pugwash Conferences, The U.S.-Soviet Study Group and the ABM treaty. Natural scientists as political actors: Historical successes and lessons for the future*. PRIF Reports No. 44. Frankfurt am Main: PRIF.

Leach, Gerald. 1965. Defeat for WHO. *The New Statesman* 25: 996.

Long, Franklin D. 1972. Eugene Rabinowitch. An appreciation. *Bulletin of the American Academy of Arts and Sciences,* 26 (1): 12–18.

Lüscher, Fabian. 2020. Party, peers, publicity. Overlapping loyalties in early Soviet Pugwash, 1955–1960. In *Science, (anti-)communism and diplomacy: The Pugwash Conferences on Science and World Affairs in the early Cold War*, eds. Alison Kraft, and Carola Sachse, 121–155. Leiden: Brill.

McElheny, Victor K. 1965. WHO shelves Research Center plan. *Science* 148: 1576–1578.

Rabinowitch, Eugene. 1963. Pugwash-COSWA: International conversations. *Bulletin of the Atomic Scientists* 19 (6): 7–12.

Rabinowitch, Eugene. 1965. About Pugwash. *Bulletin of the Atomic Scientists* 21 (4): 9–15.

Revelle, Roger. 1961. *Proceedings of the 7th Pugwash Conference, Stowe, September 1961*, 49–58.

Röhrlich, Elizsbeth. 2018. An attitude of caution: The IAEA, the UN, and the 1958 Pugwash conference in Austria. *Journal of Cold War Studies* 20 (1): 31–57.

Röling, Bernard V.A. 1961. *Proceedings of the 7th Pugwash Conference, Stowe, September 1961*, 73–77.

Rose, Hilary. 1965. Failure of the WHO Research Center. *Minerva* 5: 340–356.

Rotblat, Joseph. 1961a. Address of secretary general. In *Proceedings of the 7th Pugwash Conference, Stowe, September 1961*, 39–48.

Rotblat, Joseph. 1961b. Address of the Secretary General. In *Proceedings of the 8th Pugwash Conference, Stowe, September 1961*.

Rotblat, Joseph. 1967. *Pugwash—A history of the Conferences on Science and World Affairs*. Prague: Czechoslovak Academy of Sciences.

Rotblat, Joseph. 1972. *Scientists and the quest for peace. A history of the Pugwash Conferences*. Cambridge: MIT Press.

Rubinson, Paul. 2020. American scientists in "Communist Conclaves." Pugwash and anti-communism in the US. In *Science, (anti-)communism and diplomacy: the Pugwash Conferences on Science and World Affairs in the early Cold War*, eds. Alison Kraft, and Carola Sachse, 156–189. Leiden: Brill.

Slaney, Patrick D. 2012. Eugene Rabinowitch, the Bulletin of the Atomic Scientists, and the nature of scientific internationalism in the early Cold War. *Historical Studies in the Natural Sciences* 42 (2): 114–142.

Topchiev, Aleksandr. 1961a. International scientific cooperation and the prospects for its development. In *Proceedings of the 7th Pugwash Conference, Stowe, September 1961*, 29–38.

Topchiev, Aleksandr. 1961b. On general and complete disarmament. In *Proceedings of the 8th Pugwash Conference, Stowe, September 1961*, 193–204.

Chapter 4
The Pugwash Leadership: Informal Diplomacy Beyond the Conferences, 1962–1967

Abstract This chapter focuses on a new 'targeted' mode of informal diplomacy carried out by the Pugwash Continuing Committee which involved informal 'back-channel' interventions beyond the conferences. Three episodes are examined, beginning in October 1962, when Pugwash scientists sought to provide a mediation channel between Moscow and Washington during the Cuban missile crisis. The second intervention in Spring 1963 involved American, Soviet and British scientists collaborating on a technical solution ('Black box' unmanned seismic detection technology) to problems surrounding the 'detection and inspection' of nuclear tests, which constituted major stumbling blocks in the Limited Test Ban Treaty (LTBT) talks in Geneva. A technical report by Pugwash was subsequently taken up within the official LTBT negotiations process. The third intervention came in 1967 in the context of the Vietnam War, when the Pugwash leadership, notwithstanding deep internal divisions along bloc lines over Vietnam, launched an audacious mission in shuttle diplomacy between Hanoi and Washington to try to end the conflict.

Keywords Cuban missile crisis · LTBT · Black box · Vietnam war · Back-channel diplomacy · Lev Artsimovitch · Bernard T. Feld · Henry Kissinger · Herbert Marcovich · Mikhail M. Millionshchikov · Joseph Rotblat

The afterglow of the London conference was eclipsed by another twist in the Cold War when, in October 1962, the Cuban crisis suddenly brought the superpowers to the nuclear abyss. Rotblat and his colleagues on the Continuing Committee seized the moment, mobilizing rapidly into what was in effect an East–West alliance of scientists, seeking to mediate between Moscow and Washington. This was an indication of the growing confidence of the Pugwash leadership in its capacity to act on the international stage based both on their credentials as scientists and as leaders of the Pugwash organization. With this highly improvised initiative, the Pugwash leadership moved into uncharted territory, developing a different mode of informal diplomacy beyond the conference setting.

The Cuban intervention set a new precedent. This chapter describes two further targeted 'behind the scenes' interventions by the Pugwash leadership undertaken in the 1960s during periods of heightened Cold War tension. In 1963, Pugwash scientists made a technical contribution towards resolving the linked problems of 'detection and inspection' that constituted a major stumbling block in negotiations in Geneva on the Limited Test Ban Treaty. The second part of this chapter examines a remarkable episode that began in June 1967 in which Pugwash scientists engaged in a form of 'shuttle diplomacy' between Hanoi and Washington as they sought to help to bring about an end to the war in Vietnam. This is contextualized beforehand with a review of the tensions along bloc lines created by this conflict within Pugwash, evidenced most strikingly at the conferences in Venice and Sopot in 1965 and 1966 respectively. The Vietnam intervention widened the organization's role in conflict moderation, in terms both of moving beyond an immediate concern with nuclear weapons and in that it involved Pugwash scientists enrolling non-Pugwash colleagues into their plan. The leadership mounted this foray into 'shuttle diplomacy' at a time when the Vietnam War was tearing the PCSWA apart. Yet this crisis—marked by open hostilities at conferences—did not preclude cross bloc cooperation between Pugwash scientists in private in order to mount a daring back-channel diplomatic mission as messengers between Hanoi and Washington.

4.1   October 1962: The Cuban Missile Crisis

At the height of the Cuban crisis in late October 1962, Pugwash veterans Harrison Brown and Joseph Rotblat were constantly in touch with each other, convinced that the organization could and should do something.[1] Based on consultation with his American colleagues close to the Kennedy administration who had links to Pugwash—unnamed, but likely members of the East Coast elite—Brown sent a cable, via the Secretary General, to Academician Aleksandr V. Topchiev, chair of the Soviet Pugwash group, urging him to try to influence the Kremlin to act with restraint. Brown assured Topchiev that likewise he was urging Washington to do the same. Rotblat subsequently proposed a meeting of senior Soviet and American Pugwash scientists to discuss the crisis and seek a possible solution. The terms of reference for the meeting were agreed by telephone with their respective governments in Moscow, Washington and London—where the meeting was to take place. During this period, Rotblat described himself as having "for several days" been in "almost continuous telephone communication with Washington and Moscow"—although it remains unclear as to who exactly he was speaking with (Rotblat 1967, 58). In the event, however, on Sunday 29 October, the Cuban crisis abated, averting the need for action.

---

[1] Johan Galtung. 'The role of Pugwash as a communication link between Washington and Hanoi.' Report dated 12 September 1968. RTBT 5/3/1/22-2.

Although nothing concrete came of the efforts launched by Pugwash scientists, they had been able to establish a cross-bloc channel of communication in a time of acute crisis between the superpowers, and had agreed plans to hold a mediation meeting. In this, they had received tacit support at the highest levels in Moscow and Washington, suggesting acceptance in these circles of Pugwash scientists as 'back-channel' intermediaries with a role to play in conflict moderation. Certainly, the small cohort of Pugwash scientists involved in the Cuban intervention took confidence from this first foray into informal diplomacy on the wider international stage. For them, this experience was a legitimation of Pugwash acting in an expanded 'back-channel' role beyond the conferences—and a platform on which to build.

At the next (sixteenth) meeting of the Continuing Committee in February 1963, the Secretary General reported very briefly on these events.[2] This low-key coverage of what those present knew to have been a major step for Pugwash, in terms of the scope of its role on the international stage and its profile in Moscow and Washington, set a precedent for how its unofficial 'back channel' role was to be handled in the internal records of the organization. Moreover, Pugwash scientists beyond the Continuing Committee knew little of the detail of such activities, typically learning of them in retrospect.

Meanwhile, the composition of the Continuing Committee was beginning to change in ways that reflected the growing influence of Europeans within the organization. At the London conference in 1962 the decision was taken to expand the Committee to include Western Europeans and also the Indian physicist, Vikram Sarabhai (Kraft 2020). However, the minutes of its next meeting, in February 1963, indicate that only scientists from the US, the Soviet Union and the UK were present. At this meeting, a new Executive Committee was created specifically to facilitate initiatives such as that taken during the Cuban crisis. That is to say, this Committee was to coordinate the new 'extra-Conference' mode of informal diplomacy: its founding members were Joseph Rotblat (as Secretary General, holding a permanent seat), Bentley Glass, Cecil Powell, Dimitrii Skobel'tsyn, and, for a year, taking a seat that was placed on an annual rotating basis, the French physicist Herbert Marcovich. In effect, this ensured that back-channel diplomacy beyond the conferences—carried out under the Pugwash banner—remained largely in the hands of American, Soviet and British scientists. The timing of the creation of the Executive Committee is noteworthy: it came into being just prior to the expansion—in practice—of the Continuing Committee. Its next (seventeenth) meeting held in September 1963 was the first to involve western Europeans and Vikram Sarabhai.[3] Seen in this light, the Executive Committee can be interpreted as a means by which power within Pugwash remained concentrated in British, American and Soviet hands.

---

[2] Minutes of the sixteenth meeting of the Continuing Committee, 8–10 February 1963, the Ciba Foundation, London. RTBT 5/3/1/2 (Pt 1) (4).

[3] Minutes of the seventeenth meeting of the Continuing Committee, 17–25 September 1963, Hotel Neptun, Dubrovnik. RTBT 5/3/1/2 (Pt 1) (4).

The new Executive Committee enjoyed enormous discretionary powers. It was able to make decisions independent of the full Continuing Committee, including regarding when and how to launch informal back-channel initiatives, activating networks and deploying Pugwash resources as/when it saw fit—acting always under the auspices of the (wider) PCSWA. These interventions typically began with telephone calls or cables, and a small private meeting, lasting two days or so, where those present decided upon a course of action. At the time, these meetings, almost always coordinated by Joseph Rotblat, were handled privately—certainly not advertised and even secret, although senior Pugwash scientists disliked talk of secrecy. An internal hierarchy was taking shape and, in a strengthening pattern, those at the apex of the organization were asserting a new degree of autonomy and agency, not least as they branched out into developing new modes of informal diplomacy.

In March 1963, a month after it had been formed, the Executive Committee mounted another very different informal intervention, one focused on a stumbling block in the LTBT negotiations in Geneva. On this occasion, the scientific and technical expertise of Pugwash scientists was to the fore.

4.2   Towards a Technical Solution: Pugwash Scientists and a Novel Approach to the Detection of Nuclear Tests

The radioactive dangers that nuclear weapons tests posed to human health globally had been a driving force behind the inception in 1957 of the PCSWA and remained high on its agenda during its early years (Kraft 2018). In 1958, the Vienna Declaration had called for an end to nuclear tests, highlighting their radiological dangers. In November 1958, partly in response to mounting anti-nuclear protests around the world, the three nuclear powers (the US, USSR and the UK) agreed a moratorium on above ground testing. This remained in place until 1 September 1961 when, as discussed in Chap. 2, the Soviet Union restarted weapons testing, with the Americans following suit in April 1962. A great deal was at stake when it came to weapons testing. Essentially each test—typically conducted in series—was a vast and complex technoscientific experiment generating new data and technical insights invaluable to weapons design. Nuclear weapons tests were therefore an engine for the arms race— hence the reluctance to end them, the secrecy surrounding them and why stopping them became coupled to disarmament. Negotiations towards the LTBT formed part of the on-going disarmament talks in Geneva, where the detection of tests and the inspection of test sites were major stumbling blocks.

The issue of nuclear weapons testing straddled the boundary between science and politics, and finding agreement on how to reduce and/or bring national testing programs to a halt was laden with scientific and technical questions. To the scientists of Pugwash, their expertise was directly relevant to solving the seemingly intractable technoscientific problems of test detection, inspection and monitoring that had long bedeviled international discussions about a nuclear test ban—in terms of both Comprehensive- and Limited- Test Bans: here they perceived an opportunity to contribute to the disarmament process (Barth 1998). The LTBT specifically prohibited atmospheric tests but not underground testing, which assumed new importance. Test detection by seismic means moved to the fore in order, not least, to monitor underground testing, which was seen as important in deterring secret weapons testing and development (Barth 1998).

The detection of tests and the on-site inspection of test sites became intractable problems in the LTBT negotiations which could not move forward without agreement on an international framework for both. The US and the Soviets were diametrically opposed on the question of on-site inspection. The Washington position emphasized the need for international seismic stations with obligatory on-site inspections by an International Control Commission. By contrast, Soviet proposals stressed national seismic stations (at test sites) with international control of seismic data, without on-site inspections. Accordingly, the practice and number of on-site inspections constituted a major sticking point in the LTBT process.

In tackling this problem, the PCSWA adopted a proposal by the eight neutral countries in Geneva which advocated "the utilization of a world-wide net of seismic stations manned by nationals of the host country which will provide an International Control Commission with seismic records." It was here that Pugwash scientists conceived a new approach based on the use of automatic (unmanned) seismic recording stations. This new technology could, Pugwash scientists argued, minimize disruption to the host country, "substantially reduce" the number of on-site inspections and open a pathway to compromise in Geneva.

The seed of this idea was planted in cross-bloc discussions at the ninth conference in Cambridge, UK, in August 1962. The key figures included from the Soviet Union Lev Artsimovitch, J. Riznichenko and Igor Tamm, and the Americans David R. Inglis, Richard S. Leghorn and Alexander Rich. Based on the concept of "sealed automatic seismographs," this group put forward the idea of installing unmanned automatic seismic stations that included an automated reporting system—so-called "Black boxes"—in nuclear testing zones.[4] This new "Black box" technology afforded the technical means to identify, detect and record seismic data about nuclear tests, thereby potentially providing a means to reduce the number of on-site inspections. Although the "Black box" concept did not resolve questions about the exact number of inspections— in principle, anathema to the Soviets—it raised hopes of a compromise solution that, in turn, could break the deadlock in Geneva. (This technology also offered, potentially, the means to distinguish seismically between nuclear explosions and earthquakes, a critical distinction in terms of putting in place an international system for monitoring tests, including underground.)

Plans to take this idea forward were derailed by the sudden death of Aleksandr Topchiev in late December 1962.[5] Topchiev's sudden death was a huge blow to Pugwash, and it threw the leadership into disarray. Close to Khrushchev, Topchiev

---

[4] This basic idea was provisionally set out in: 'A new approach to the test ban negotiations.' Tenth PCSWA, London, 1962. RTBT 5/3/1/12 (2). Rotblat, Joseph. Pugwash and the Test Ban, 1. RTBT 5/3/1/22-2.

[5] Minutes of the sixteenth meeting of the Continuing Committee, 8–10 February 1963, the Ciba Foundation, London. RTBT 5/3/1/2 (Pt 1) (4).

was highly respected by his colleagues on the Continuing Committee and was regarded as having played a key role in enabling and maintaining the cross-bloc character of both the Committee and the wider organization (Lüscher 2020). His interim replacement on the Committee was Vladimir Kirillin, who was replaced in 1964 by Mikhail D. Millionshchikov, when Lev Artsimovitch also became a member (Kadomtsev 1985).[6]

In February 1963, Joseph Rotblat and Harrison Brown moved to restart the Black box project. In a cable to Brown on 25 February, Rotblat reported "considerable interest here" in holding a small East–West meeting in the next two weeks to set the ball rolling again and involving those "intimately knowledgeable" and emphasizing "there should be no publicity."[7] The source of the (British) "interest" to which Rotblat referred was the seismologist Edward ("Teddy") Bullard who was well-connected in Whitehall. On the same day, the two had discussed the situation by telephone, during which Bullard told Rotblat that he had written to Solly Zuckerman and William Penney about the "Black box" idea.[8] A small private meeting for discussions between British, American and Soviet scientists about this idea was promptly scheduled for 16–18 March 1963 which, by arrangement with the director, Gordon E. W. Wolstenholme, took place in London at the Ciba Foundation (Wolstenholme 1949).[9] Pugwash networks were activated and mobilized. At very short notice, twelve scientists—from the US, Soviet Union and the UK, including seismologists—all of whom were close to their respective governments—came together for strictly confidential talks about the innovative "Black box" proposal.

Chaired by Joseph Rotblat, the agenda focused on the technical aspects of detecting underground tests, the "political issues" of on-site inspection and the "prospects for reaching agreement on a comprehensive test ban." Those present were, from the US, Pugwash veterans Harrison Brown, Paul Doty, and Alexander Rich, the chemist George Kistiakowsky (1900–1982), the physicist Matthew Sands (1919–2014) and the geophysicist Frank Press (1924–2020).[10] The three Soviet

---

[6] Minutes of the nineteenth meeting of the Continuing Committee, September 1964, Prague. RTBT 5/3/1/2 (Pt1) (4).

[7] Joseph Rotblat to Harrison Brown, Western Union Cablegram, 25 February 1963. RTBT 5/2/1/11 (30).

[8] Edward Bullard to Joseph Rotblat, 27 March 1963. RTBT 5/2/1/11 (30). Bullard was reluctant to become involved but, in the event, attended the Ciba meeting for two days.

[9] The Ciba Foundation was an educational and scientific charity funded by the Swiss chemical company, Ciba-Geigy. Its offices in London on occasion played host to the meetings of the Pugwash Continuing Committee. For example: 16th Meeting, 8–10 February 1963. RTBT 5/3/1/2 (Part 1) (4). This was made possible by the director of the Ciba Foundation, Gordon Wolstenholme (1913–2004), a medical doctor with an army background, who served as President of the Royal Society of Medicine between 1975 and 1977. Wolstenholme occasionally attended Pugwash conferences, including that held in Addis Ababa in 1966, reflecting his particular interest in Pugwash engagement with the countries of the Global South. Other venues for meetings of the Continuing Committee in the English capital included the Russell Hotel in Bloomsbury, and St. Bartholomew's Hospital, where Rotblat was professor of physics.

[10] Matthew Sands was a consultant to the PSAC on disarmament. The geophysicist Frank Press was of Russian émigré background and was an early member of the PSAC; later he served as science

participants were Lev Artsimovitch (1909–1973), the seismologist Riznichenko, and Vladimir P. Pavlichenko. In addition to Teddy Bullard, the British contingent included Drs. M. Hill and T. Armstrong—all geology/seismology experts.[11] Pragmatically, all conceded that "there was not much point" in pressing for a comprehensive test ban, but there was agreement that a partial test ban was feasible—to which the Russians had initially been opposed.[12] The meeting focused on developing mathematical models relating to underground testing which extended into "the calculation of the probability of detecting a series of underground tests."[13] Working together, they produced a short technical report on the Black box concept and its application to the detection and inspection problem. This included tables of probability calculations on test detection, with/without "Black boxes," and addressed the mathematical difficulties of distinguishing between earthquakes and tests.[14] All agreed to pass along the report to their respective governments.[15] This Pugwash intervention constituted a two-stage process of informal diplomacy. The first techno-scientific stage centered on private exchanges between scientists at the Ciba Foundation in London which resulted in a technical report. The second stage comprised the journey of this report as it travelled into and within government circles, for example, in the British case, via Solly Zuckerman, and from there to the negotiating table in Geneva.

Edward Bullard had missed the final day of the meeting, but Rotblat later updated him that discussions had been "quite successful" following an unexplained shift in the stance of the Russians, speculating that "whether (they) had second thoughts…or had some link with Moscow, but they were certainly much more forthcoming."[16] Rotblat was of the view that the technical report drawn up during the meeting "gives promise of compromise." He also relayed to Bullard that George Kistiakowsky had telephoned Lev Artsimovitch on 20 March to tell him "that the Americans are pleased with the results of the meeting" and moreover, that Artsimovitch had indicated that the Russians "too are optimistic." Concluding that "It may turn out that this meeting was not in vain," Rotblat reminded Bullard to treat the document as "confidential" emphasizing that "both the Americans and the Russians were very keen that it should not receive any publicity." Rotblat circulated a report about the meeting to the Continuing Committee and sent a copy to Solly Zuckerman; a distribution list in Rotblat's

---

advisor to President Jimmy Carter and between 1977 and 1981 was director of the Office of Science and Technology Policy.

[11] Bullard, along with Doty, Kistiakowsky and Press, were able to attend only the first two days of the meeting. RTBT 5/2/1/11-30.

[12] The report summarized that "the whole problem of on-site inspection was really a political issue which could not easily be solved in the political climate of the time." Galtung, Johan. Pugwash as a Communication Channel in Times of Crisis, 5. RTBT 5/3/1/22-2.

[13] Nuclear tests were not carried out singly, but typically in 'series' involving multiple tests—each of which was a 'stand-alone' experiment investigating a particular aspect of, for example, bomb design or performance.

[14] 'Short technical report.' March 1963. RTBT 5/2/1/11 (30).

[15] A copy of this document was enclosed in a letter from Rotblat to Solly Zuckerman, 27 March 1963. RTBT 5/2/1/11 (30).

[16] Joseph Rotblat to Edward Bullard, 22 March 1963. RTBT 5/2/1/11 (30).

hand indicates that he had also sent copies to the British Prime Minister, Harold Macmillan and to Premier Khrushchev.[17]

Rotblat also reported privately to the Norwegian sociologist and peace activist Johan Vilhelm Aubert his view that the meeting had been "very successful. The attendance was very high-powered and really got down to a detailed analysis of the technical aspects of inspections and political difficulties in reaching agreement."[18] He went on to say that document drawn up "shows that both sides have agreed to influence their governments to reach a compromise" and that here "specific proposals" were agreed which had "already been submitted to the governments concerned." Rotblat closed by expressing hopes that in the coming weeks this might lead to results, adding that it was "very important at this stage no publicity be given to the meeting" and, to this end, asked Aubert to treat the information as "confidential."

Subsequent reports indicate that the Black box concept was "taken up by the Soviet government, both in a letter from Khrushchev to Kennedy and in official negotiations;" moreover, Jerome Wiesner later directly linked the signing of the LTBT in Moscow in August 1963 to the Ciba Pugwash meeting (Rabinowitch 1963, 9).[19] At the very least, the concept of Black box technology fed into conversations in Geneva in which the detection/inspection problem was resolved, paving the way for the LTBT (Kubbig 1996; Barth 2006). This 'unofficial' contribution was made possible by the scientific and technical expertise of scientists from East and West brought together by and cooperating under the aegis of the Pugwash organization. In hindsight, this stands as a unique demonstration of the purpose and the value of the Pugwash project—one that embodied its founding vision of scientists transcending the bloc divide and contributing to disarmament. At the same time, it provides an example of complementarity between unofficial and official diplomacy.

Within Pugwash, all of this was handled in a low-key way, and recorded very briefly in the minutes of the next meeting of the Continuing Committee in September 1963.[20] Under the heading "Test Ban meeting," it was stated simply that the Ciba meeting had been organized by the Secretary General after "consultation with the Executive Committee," and was fully endorsed by the Continuing Committee. It reported that the discussions had been "very fruitful and the participants agreed to convey the findings to their governments." The understated and cursory treatment of this intervention within the official internal records of the PCSWA reflected the by now familiar strategy on the part of the leadership of downplaying this kind of

[17] Rotblat, Handwritten list. RTBT 5/2/1/11 (30). Whether the report reached Macmillan or Khrushchev remains unclear.

[18] Rotblat to V. Aubert, 4 April 1963. RTBT 5/2/1/11 (21).

[19] Johan Galtung. 'The role of Pugwash as a communication link between Washington and Hanoi.' Report dated 12 September 1968. RTBT 5/3/1/22-2. Seemingly, ahead of the Ciba meeting, the Black box project was the subject of controversy, specifically with disagreements between Rotblat and Henry Kissinger aired in the letters pages of the *New York Times* on 10 January 1963; 26 February 1963. The background to this unusual situation remains unclear, but certainly it ran contrary to the Pugwash preference to keep its role in informal diplomacy out of the spotlight.

[20] Minutes of the seventeenth meeting of the Continuing Committee, 17–19 and 24–25 September 1963, Neptune Hotel, Dubrovnik, 2. RTBT 5/3/1/2 (Pt. 1) (4).

work. As in official diplomacy, discretion was essential: such practices were about protecting the organization's back-channel role so that it might continue.

The PCSWA had been formed in response to the advent of the thermonuclear age in the mid-1950s Cold War—but by the early-mid 1960s the contours of the conflict were changing rapidly. These changes profoundly affected the internal dynamics of the organization, including relations within the Continuing Committee, and between this Committee and colleagues within the wider organization. Amongst the most serious challenges confronting Pugwash in this period was the war in Vietnam: this destabilized its leadership, led to open hostilities along bloc lines at conferences, and damaged the standing of the international PCSWA in Washington. Indeed, in the mid-1960s, the conferences became a bitter Cold War battleground as Soviet scientists repeatedly condemned the Johnson administration for its aggression against its communist ally, North Vietnam. Paradoxically, for all these deep internal divisions, in June 1967 the Vietnam War provided the context in which the Pugwash leadership launched a third—and perhaps its most audacious—initiative to date in the realm of informal diplomacy.

4.3   The Vietnam War: The Reality of Bloc Allegiances Within Pugwash

During the mid-1960s, as the superpowers moved fitfully towards a period of nuclear détente, signaled by the LTBT and talks that in 1968 resulted in the Non-Proliferation Treaty (NPT), the Vietnam War became a violent flashpoint between them. In protest the Soviet leadership held to a policy of blocking official contacts with Washington—a "blockade" which also seriously affected the work of the Soviet American Disarmament Study Group (SADS), stalling talks on Anti-Ballistic Missiles (ABM) for much of the period between 1965 and 1967 (Kubbig 1996, 30–36). During this time, the Pugwash organization and its conferences provided a rare channel of communication between Moscow and Washington. But the conferences also provided an international forum in which the Soviets could vent their anger over Vietnam: in particular, the fourteenth and sixteenth conferences in Venice in 1965 and in Sopot in 1966 descended into open and bitter hostility.

The Vietnam War brutally exposed and tested to the limits the always fragile East–West fabric of the Pugwash organization. In Venice and Sopot, scientists from both superpowers—sometimes aided by colleagues from within their respective alliance systems—traded verbal blows over Vietnam, with the Soviets on the attack, the Americans on the defense.[21] Soviet scientists—including powerful members of the Continuing Committee, especially Mikhail Millionshchikov—repeatedly launched blistering attacks on the US, especially over its use of chemical weapons against civilians. Wary of antagonizing Washington, American scientists—and their

---

[21] For example: Theodor Němec. Pugwash and Vietnam. RTBT 5/2/1/16 (8).

British colleagues—adopted a strategy of neither defending nor criticizing Johnson's policy—choosing instead to emphasize the dangers of „local wars" everywhere whilst calling for both sides to begin negotiations to end the conflict. For the western half of Pugwash, adhering to the principle of political neutrality had never been so vital. For the Eastern half of the organization, this principle had never been so weak and frustrating. These polarized positions along bloc lines exposed the faultline running through the PCSWA, laying bare one of the founding Pugwash myths that its scientists were rational, impartial actors able always to suspend and overcome national loyalties and bloc allegiances. By mid-decade, Vietnam was threatening to tear Pugwash apart.

In the run up to the fourteenth conference held in Venice in April 1965, Bernard T. Feld and Eugene Rabinowitch, leading members of the American Pugwash group and serving members of the Continuing Committee, already viewed this conference as a stern test for the PCSWA in terms of its relations with Washington (Rubinson 2020, 170–174). This was because already in Fall 1964 trenchant criticisms of American foreign policy in Central Europe had been made by European scientists at the thirteenth Pugwash conference in Karlovy Vary. This had elicited great anger in Washington, damaging relations with the Johnson White House—discussed further in the following chapter. In the event, however, it was a different aspect of American foreign policy, namely the war in Vietnam, that exploded onto the scene in Venice, dominating the discussion and laying waste to the planned program.

4.4 Venice 1965/Sopot, 1966: Pugwash Conferences as Cold War Battleground

Assembling in Venice, Pugwash scientists were acutely aware of developments in South East Asia. The conference theme was 'International cooperation for science and disarmament' and in his opening address Rotblat foregrounded the organization's contributions to disarmament, including its introduction and development of the concepts of minimum deterrence, the nuclear umbrella, the staging of the disarmament process, zonal schemes of inspection and Black box technology for the detection of nuclear weapons tests (Rotblat 1965, 72). But, acknowledging the issue on everyone's mind, Rotblat went on to say "We cannot ignore the fact that a dangerous situation now exists in Vietnam." He then provided a strong steer as to how this was to be handled, emphasizing that this was to be discussed "within the context of the main theme of disarmament" (Rotblat 1965). This guidance fell on deaf ears.

Leading the charge in Venice were the Soviet scientists Mikhail Millionshchikov—since 1964 a member of the Continuing Committee—his colleague, Mikhail Dubinin, and Theodor Němec, a leading figure in the Czechoslovakian Pugwash group. The trio demanded an immediate halt to the war, and vehemently condemned the use of chemical weapons by the US. This display of Eastern bloc solidarity reflected the position of the Czechoslovak group as the primary Eastern bloc ally of Moscow within the PCSWA. As Doubravka Olšáková has shown, this status was part of an arrangement that was far from stable, reflecting the uneasy relations between Moscow and Prague (Olšáková 2020). In 1968, the dramatic events in Prague saw the Czech group's special position vis-à-vis Moscow pass to the Polish Pugwash group. But at this moment, the Soviet-Czechoslovak alliance proved formidable in pushing the Vietnam War onto the conference floor and onto the Pugwash agenda.

The Venice program featured the familiar mix of plenary sessions and Working Groups; in addition, there was a short report on the recent meeting of the Biological Warfare Study Group in Geneva.[22] Millionshchikov used his plenary paper, ostensibly given to the theme 'Disarmament and international cooperation,' to launch the opening attack on the US (Millionshchikov 1965). As events unfolded in Venice, and to try to ensure that Vietnam did not completely dominate the conference, impromptu measures were hurriedly put in place. Vietnam was added to the agenda of two Working Groups—4 and 5—otherwise dealing with "problems" of arms control and disarmament respectively. Containing discussion of controversial topics within Working Groups became a key strategy of the Pugwash leadership as it grappled with the challenges arising from a new phase in the Cold War.

Vietnam continued to dominate the second plenary session, chaired by Rudolf Peierls—indicative of how amid the crisis the British played a mediation role— which ended in deadlock and disarray. Millionshchikov then announced that the main program would now include an additional section (Part V) given to short statements on the Vietnam situation by national groups.[23] The process leading to this unusual arrangement remains unclear: it is not referred to in either the conference *Proceedings* or the minutes of relevant meetings of the Continuing Committee—an example of the silence of official Pugwash records when it came to internal conflict and crisis. The creation within a conference of a platform specifically for expressing 'national' positions was exceptional. At the same time, this improvised mechanism served to place distance between public criticisms of American military aggression at the conference and both the western members of the Continuing Committee and the international PCSWA. In addition to further fiercely condemnatory statements from the Soviets and the Czechs, Part V included criticisms of the US from the Japanese and Yugoslavian Pugwash groups. The requisite need for balance came in the form of three individual statements from senior western scientists—Bernard T. Feld, Hermann Bondi and the American human rights lawyer Louis B. Sohn. Feld and Bondi emphasized regrets but offered no criticism of the US, each emphasizing instead the need for negotiations to halt the violence, whilst Sohn called on the UN to take a leading role in settling the crisis.[24]

Part V pointed to an organization internally convulsed by the Vietnam conflict and underlined the dangers this posed to it. However, in the Pugwash tradition, and for its international audience, the conference statement smoothed over the difficulties in Venice and foregrounded the more positive aspects of the meeting, including

---

[22] *Proceedings of the 14th Pugwash Conference, Venice, April 1965*: 253–307.

[23] *Proceedings of the 14th Pugwash Conference, Venice, April 1965*: 76–78.

[24] *Proceedings of the 14th Pugwash Conference, Venice, April 1965*: 241–250.

productive discussions in some of the Working Groups on scientific cooperation and disarmament. A short summary of 'Vietnam' was included in section 'G' of the statement which, without mentioning the US, simply acknowledged that different views had been expressed on the conflict, that the recent escalation of the conflict threatened world peace, that the UN should do all it could to end the violence, and that Pugwash scientists had agreed to pass along the different views to their respective governments.[25] The following section, 'H,' condemned the use of gas warfare anywhere in the world, but did not mention the use of these weapons by the US in Vietnam.[26]

Alarmed by what had happened in Venice, at the next meeting of the Continuing Committee in London in August, the Secretary General convened "a private meeting of scientists (members of the Committee) to exchange information about the attitudes in their countries towards the Vietnam problem."[27] No further details were included—but likely Rotblat, as Secretary General, was trying to mediate between his American and Soviet colleagues. This highlights a critical point, namely, that meetings of the Continuing Committee could, as/when needed, move beyond routine procedural matters, including serving as occasions for its members to thrash out differences in private. Rotblat and Feld were extremely worried about what might happen at the upcoming conference scheduled to take place in the Eastern Bloc, in Sopot, Poland, in September 1966. Their concerns deepened with the escalation of American bombing that summer, as Rotblat confided to Feld, "Until a few weeks ago, I had great hopes of achieving something really useful" at Sopot, explaining that "It looked then as though the détente would enable us at long last to make some real progress on specific measures on disarmament."[28] He went on:

> Unfortunately, the extension of the bombing in North Vietnam has completely changed the situation. It seems to me now very unlikely that we shall be able to shake off the Vietnam ghost from our Conference. I cannot see how we shall be able to avoid discussing this issue, and if so we should best try to channel it into a working group which would discuss the issue objectively and not for the purpose of producing a resolution and, moreover, should not begin its work until the last two days of the conference—so as not to upset ordinary work of the conference.

In reply, Feld shared with Rotblat the view that "there is little chance of excluding this 'ghost' from Sopot" and that he wanted to discuss with his American colleagues the idea of placing Vietnam into a working group. He fully anticipated that "There will of course be a strong desire on the part of many if not most of the participants to condemn the bombings" and that this would elicit a range of responses from the Americans, "from unhappy and pained deferral to strong objections" to arguments that this was lay beyond the purview of the Conference. American scientists—including within the PCSWA—held different views on Vietnam (Bridger 2015,

---

[25] Conference statement, *Proceedings of the 14th Pugwash Conference, Venice, April 1965*: 9–19.

[26] Conference statement, *Proceedings of the 14th Pugwash Conference, Venice, April 1965*: 17.

[27] Minutes of the twenty-second meeting of the Continuing Committee, 30–31 August 1965, Ciba Foundation, London, 3. RTBT 5/3/172 (5).

[28] Joseph Rotblat to Bernard T. Feld, 25 July 1966. RTBT 5/2/1/16 (33).

115–154). Privately, Feld was dismayed by the recent escalation in bombing which "was a tragically dangerous and irresponsible action which has seriously increased the danger that the war may spread." He suggested that discussion of the conflict ought to be guided along two lines, firstly, what could be done to stop the conflict spreading, and secondly, placing emphasis on "bringing all the interested parties to the negotiating table." Feld hoped also to somehow avoid the "divisive process" of agreeing on a public statement from Sopot, going on to suggest trying to arrange a meeting of a small group just prior to Sopot in Warsaw "to try to work out, if possible, a statement of position and objectives." This might be one means to "avoid any informal discussion of this issue on the conference proper" and he was keen to have at the meeting the Indian physicist Vikram Sarabhai, a serving member of the Continuing Committee, someone from Poland and from France and, if possible, the UN director, U. Thant.[29] Feld acknowledged that this too came with risks, as he put it, "Of course, we run the danger that if such a meeting can be arranged, they may conclude that the whole thing is hopeless and we will be then right back where we started; but I assume there is some reasonable way for preventing the Conference from falling apart on this issue." In August 1966, a month before the Sopot conference, Rotblat wrote to Theodor Němec—whom he knew well—suggesting that Vietnam be discussed within the Continuing Committee because he "would not like a repetition of the happening at the Venice conference when this issue dominated the whole meeting without producing any definite result."[30]

In the event, simmering tensions over Vietnam erupted again in Sopot. The conference theme—'Disarmament and world security, particularly in Europe'—resonated strongly with the interests of Europeans, who at this point were driving engagement with this topic within a newly created Study Group on European security. But the long shadow of Vietnam loomed everywhere. In his opening address as Secretary General, Rotblat acknowledged that the conflict "must be uppermost in our minds," given the intensification of the conflict, the growing danger of escalation within/beyond the region, and the "terrible suffering of the Vietnamese people."[31] Moving to contain discussion of Vietnam along the lines discussed with Feld, Rotblat added that exactly because the conflict elicited such strong feelings, "we must be careful not to allow this to preoccupy us so much as to reduce the possibilities of fruitful discussions on the other problems on the agenda." He informed the conference that, the Continuing Committee had decided Vietnam would be taken up exclusively in Working Group 3 given to the theme 'Main problems of progress towards GCD,' within a sub-section given to 'Current conflicts.' Rotblat's informal but strong steer as to how he would like to see the conference unfold points to the internal power dynamic within the PCSWA and to the existence of an unspoken code of conduct that its scientists were expected to abide by.

---

[29] A pre-meeting in Sopot did take place between 13 and 15 June 1966, but the available records of this meeting focus on the German question and do not mention Vietnam. RTBT 5/2/1/26 (32).

[30] Rotblat to Němec, 15 August 1966. RTBT 5/2/1/6 (33).

[31] Rotblat, *Proceedings of the 16th Pugwash Conference, Sopot, September 1966,* 54–56, 55.

But here again Rotblat's call went largely unheeded by his Soviet colleagues. Theodor Němec also issued a one page statement expressing trenchant criticisms of the US and arguing forcefully that on this issue, Pugwash "must not remain silent" but must instead remain "faithful to its principles and tradition."[32] In their papers in Sopot, Dubinin, Millionshchikov and Aboltin all attacked the US over Vietnam (Aboltin 1966). In his paper, focused on chemical warfare, Dubinin was scathing about US use of these internationally outlawed weapons in Vietnam (Dubinin 1966). In his lengthy paper on European security, Millionshchikov pulled no punches, attacking the "ruling circles" of the US for "their unjustified, wanton war in Vietnam" (Millionshchikov 1966). Acknowledging his audience—he regretted too that "I have to say this here in the presence of our American colleagues, but the bitter truth is better than any diplomatic omissions. The US ruling circles disgrace themselves in Vietnam," and they should not be surprised that their actions "resound in Europe." He reiterated the call—made time and again by the Soviets—for the US to cease bombing North Vietnam, withdraw its forces (and those of its allies) from the region completely, and to leave the Vietnamese people to self-determination—in line with the UN Geneva Agreements of 1954. Provocatively, Millionshchikov concluded by making a direct link between US policy in Vietnam and in Europe: "The US war in Vietnam is a part of a general aggressive course pursued by the influential circles of the US. And they follow the same course in Europe, where US ruling quarters, through the Atlantic Pact and its military organization, aim at increasing international tension and widening the split on the European continent." Millionshchikov would have been well aware that this resurrected arguments made in 1964 in Karlovy Vary when criticisms by some Pugwash scientists of American policy in Central Europe had so angered Washington, especially the State Department (Rubinson 2020, 171–174).

The next paper by Bernard Feld was very different. Operating completely in conciliatory mode, Feld held assiduously to the lines he had indicated privately to Rotblat, namely, calling for restraint on both sides in Vietnam and urging both to begin negotiations. Calling also for calm within Pugwash, Feld exemplified the way in which the western leadership did all it could to maintain and assert for the organization a position of neutrality. He concluded by sharing a personal recollection, prompted by his journey through the Polish countryside to Sopot, about the experience of his parents, Polish Jews who had sixty years earlier fled the region to escape Tsarist persecution (Feld 1966). This personal sensibility to distress and the destruction wrought by political tyranny was shared by others from an émigré background— a background not uncommon within the western Pugwash leadership. For them, Pugwash afforded perhaps a means to channel revulsion and anger over past violence and injustice into a project working for peace.

But Vietnam stayed at the top of the agenda in Sopot. In an amendment to the program, an additional evening session of the second plenary session on 13 September given to the conflict was arranged specifically "to enable all members of the conference to express their views on this vital issue." Chaired by Rudolf

---

[32] Theodor Němec. Pugwash and Vietnam. RTBT 5/2/1/16 (8).

Peierls, the meeting involved Feld, Millionshchikov and Rotblat along with twenty-four colleagues including Lev Artsimovitch, Mikhail Dubinin, Vasily Emalyenov, Johan Galtung, Bentley Glass, Vladimir Khvostov, and Henry Kissinger. In the conference *Proceedings,* the report for this meeting was limited to a list of those present and a short note explaining that it had been agreed that no statement would be issued from it.[33] What transpired remains so far unknown.

Many of those taking part in this evening session were involved in Working Group 3 which gave "special and separate consideration of Vietnam." Its members included the Soviet trio Dubinin, Millionshchikov and Emalyenov and five Americans—Feld, Kissinger, Frank Long, Freeman Dyson, and Sheperd Stone from the Ford Foundation, alongside Karol Lapter from Poland, Vikram Sarabhai, and Joseph Rotblat.[34] Its report provided a brief summary of the discussions:

> There was considerable discussion about the causes of the war: some wished to condemn what they called the American aggression, others wished to condemn the attack on South Vietnam by what they called North Vietnam guerillas and main force units. All agreed that the tragic destruction of human life was the first consideration. Great dismay was expressed at the escalation of the war and in particular at the use of new types of chemical weapons including those for the destruction of agricultural crops. Some felt that Vietnam was being used as a proving ground for such weapons.[35]

Nevertheless, consensus had been reached in four areas. The group expressed "grave disquiet" about the escalation of violence in Vietnam, regretted the civilian loss of life, proposed that a solution to the conflict ought to be based on the 1954–1962 Geneva Agreements, including the withdrawal of foreign troops from South Vietnam and for the Vietnamese people to settle their own affairs, and it appealed to scientists everywhere to help resolve the conflict.[36] Although the idea was mooted about setting up a study group on Vietnam, this went no further. In its style—a detached tone and, importantly, a narrative marked by anonymity—this report exemplified how polarized discussions within Pugwash were recorded. Nor did the conference statement from Sopot dwell on the Vietnam issue, omitting details of Working Group 3. Rather, coverage was limited to a brief summary that noted, "While there was an extensive and frank exchange of views among the participants, the Conference was unable to arrive at any agreed conclusions on the causes and nature of the war and on possible ways for bringing to an end this dangerous and tragic conflict."[37] This belied the damage that the polarizing effects of the Vietnam War were having within the Pugwash organization.

---

[33] *Proceedings of the 16th Pugwash Conference, Sopot, September 1966*: 107.

[34] PCSWA 16. Report of Working Group 3, Main problems of progress towards GCD. *Proceedings of the 16th Pugwash Conference, Sopot, September 1966*: 32–34.

[35] *Proceedings of the 16th Pugwash Conference, Sopot, September 1966*: 32.

[36] *Proceedings of the 16th Pugwash Conference, Sopot, September 1966*: 33.

[37] Statement from the Sopot Conference. *Proceedings of the 16th Pugwash Conference, Sopot, September 1966*: 11–18.

Sustained and vehement criticisms at Pugwash conferences from scientists from the Eastern bloc, but also from Japan and Yugoslavia, of America's prosecution of the war in Vietnam was resented in the U.S. At the same time, Pugwash drew fire from the socialist bloc countries for what they saw as its failure to condemn sufficiently US aggression. Indeed, throughout these tense years of division and enmity, on the international stage the PCSWA maintained an official position of neutrality on the Vietnam War. Essentially, this meant lamenting civilian casualties, issuing only general statements against the use of chemical weapons without mentioning the US or Vietnam, and calling for both an immediate end to military action and for negotiations to begin. Locked in a vicious circle of accusation and recrimination in the mid-1960s, the Pugwash organization lurched from one conference to the next where the Secretary General and his closest allies struggled to mediate between their colleagues who remained bitterly divided along bloc lines. In October 1966, articles in Soviet newspapers about the Sopot Conference by Emelyanov and Millionshchikov, which implied that everyone at Sopot "strongly protested against US barbarity in Vietnam," caused great alarm within senior American Pugwash circles, who saw in them distortions of what had transpired and worried about the effects of such reports in Washington (Emelyanov 1966).[38]

The presence of the American Cold Warrior Henry Kissinger in both Venice and Sopot—and also at the preceding conferences in Dubrovnik (1963) and Karlovy Vary (1964)—raises interesting questions. Working for the American government in the realm of official diplomacy, including in matters relating to science and technology, Kissinger was a key figure in senior Washington circles (Kistiakowsky 1974, 40). Given the tensions between the PCSWA and the Johnson administration, perhaps in part his role included keeping an eye on the organization. At the very least, Pugwash conferences afforded him a rare opportunity to hear first-hand criticisms of US foreign policy from the Soviets and others. Likely too he was seeking to take advantage of the opportunities afforded by the conferences for encounters and conversations not available elsewhere in this period—implicit recognition perhaps of the conferences as an unofficial forum for informal diplomacy. The major themes of this string of conferences between 1963 and 1966 were European security and Vietnam, and Kissinger was active in discussions about both at plenary sessions, and participated in various Working Groups—including Working Group 3 in Sopot, which encompassed Vietnam, and Working Groups 1 and 3 at Karlovy Vary and Dubrovnik respectively, which dealt with European security. All three of these groups elicited a degree of controversy—and whilst privy to the discussions, Kissinger was careful to steer clear of putting his name to any statements issued from them. It is worth recalling that Working Groups operated in private, informally, and were governed by Chatham House rules. In practice, this rendered them—potentially—rich sites for informal diplomacy in providing opportunities for sensitive discussions at a sub-official level and completely out of the spotlight.

---

[38] Franklin D. Long to Joseph Rotblat, 1 November 1966. RTBT 5/2/1/16 (23). Long refers to a recent article by Millionshchikov in *Pravda*.

Indeed, perhaps this was exactly the point. Kissinger's repeated appearances at the PCSWA in this period suggests that he perceived value in them as a place where off-the-record discussions with the Soviets and their allies was possible, but also because at this time, tensions and confrontations over Vietnam at the conferences opened a rare, real-time window on to how America was perceived in both East and West. Circling elusively in/around the conferences where he exercised a unique American influence, Kissinger dipped into discussions and tapped into the organization's cross bloc networks without putting his name to anything. That said, Theodor Němec reported that late in 1964, Kissinger secured funds that supported the creation of the Pugwash Study Group on European Security.[39] For a time, Kissinger was a familiar figure at the conferences and in summer 1967, he was pivotal to an informal diplomatic initiative launched by the Pugwash leadership in Vietnam (Salomon 2001).

The angry confrontations in Venice and Sopot exposed the limitations of the Pugwash organization as an East–West bridge when faced with a violent conflict rooted in the ideological rivalry between the superpowers. At some moments, the future of the Pugwash project seemed to be in jeopardy.[40] However, for all the recent enmity at conferences, in June 1967, the Continuing and Executive Committees conceived and worked together on an unofficial diplomatic mission in Vietnam to try to end the conflict.

4.5   June–July 1967: Shuttle Diplomacy Between Hanoi and Washington

Against a backdrop of internal crisis and division, and working informally behind the scenes on the international stage, the Executive Committee sought to provide a channel of communication between Hanoi and Washington. This initiative marked a departure from the traditional terrain of Pugwash, as its scientists moved beyond an immediate concern with nuclear weapons and disarmament into a broader role in conflict moderation. It rested on a number of factors, not least its French connections.[41] Initially guided by Rotblat, the central figure in this operation was the French biologist Herbert Marcovich, a member of the Continuing Committee since 1962.[42] Henry Kissinger was pivotal, providing a direct link with the US administration. On the Soviet side, Mikhail Millionshchikov sanctioned the mission: whilst not actively involved, he was kept apprised of developments.

---

[39] Theodor Němec to Joseph Rotblat, 12 December 1964. RTBT 5/2/1/13 (43).

[40] Joseph Rotblat 1967. Memorandum: Future of Pugwash, 4. RTBT 5/3/1/19.

[41] The historiography of the Pugwash organization in France remains so far wholly under-developed.

[42] Marcovich had been appointed to the Continuing Committee in 1962 when its membership was expanded to twelve, along with the Italian physicist, Edoardo Amaldi, the Polish physicist, Leopold Infeld, and the Indian physicist Vikram Sarabhai.

When the Six-Day War broke out in the Middle East on 5 June 1967, members of the Continuing Committee immediately contacted each other—responses which as Rotblat put it, had "been the practice in periods of crisis"—in order "to determine if some action could be taken."[43] There followed a series of cables and telephone calls between Rotblat and his colleagues on the Executive Committee, Mikhail D. Millionshchikov and Franklin D. Long. But Millionshchikov was adamant that any discussion of the Middle East crisis was contingent on widening the scope of the discussion to include the Vietnam War. Millionshchikov got his way. As a first step, a small private Pugwash meeting took place in Paris between 16 and 18 June at the laboratory of physicist Francis Perrin in the College de France, and at the home of his Pugwash colleague and former resistance fighter, Etienne Bauer.[44] Present were Millionshchikov and his colleague, Academician Ruben Androssian, the Americans Paul Doty, Bernard T. Feld and Henry Kissinger, and three leading members of the French Pugwash Group, Marcovich, Bauer and Perrin: Joseph Rotblat was in the chair.[45]

During the discussions about Vietnam, the "spark" for a Pugwash plan of action was Kissinger's disclosure that the White House would stop bombing North Vietnam if the authorities in Hanoi would give assurances that they would not take this as an opportunity to increase infiltration into the south. Kissinger's involvement marked his first foray into the Vietnam situation (Kubbig 1996, 35). Galtung reported that Kissinger said "he had talked to certain people in his own government and had reasons to believe" that such a proposal would be acceptable.[46] This was henceforth adopted as a working hypothesis: in Galtung's account, the idea took shape to send to Hanoi "a Pugwashite to convey this information." Because of its past ties to Vietnam as a colonial power, it was agreed that the French had a primary role to play. A plan was hatched to dispatch Marcovich to Hanoi. Specifically, his brief was to probe how Hanoi would react to the proposal, indicated by Kissinger, to exchange a halt in the bombing for a promise "not to increase infiltration into the South, with no condition attached to it." Significantly, Millionshchikov signaled Soviet agreement

---

[43] In Rotblat's history of the early PCSWA, he refers to a small, private high-level Pugwash meeting held in London in August 1965 to discuss the Vietnam War. To date, the author has been unable to find primary sources relating to this meeting.

[44] Document listing various cables between senior Pugwash scientists about how Pugwash could respond to the Middle East crisis and the Vietnam situation, 5–14 June 1967, and an account of the Pugwash intervention in/about Vietnam in 1967–1968. RTBT 5/3/1/36 (3).

[45] Bernard Feld was on sabbatical in Paris during 1967, and Henry Kissinger was also in Paris at that time. RTBT 5/3/1/36 (3).

[46] Johan Galtung. 'The role of Pugwash as a communication link between Washington and Hanoi.' Report dated 12 September 1968. RTBT 5/3/1/22-2. Marked 'Strictly confidential. For the Nobel Committee only,' 2. The Norwegian sociologist and peace activist Johan Galtung became well known for his theorization of peace, in particular the concept of 'structural violence;' in the mid-1960s he was a regular participant at Pugwash conferences. Galtung's report was based in part on interviews undertaken at the 18th Pugwash Conference held in Nice, France, 11–16 September, 1968. This was a confidential five-page summary of this episode written specifically for the Nobel Prize Committee in support of the nomination of the PCSWA for the Nobel Peace Prize. Accordingly, Galtung's account has to be seen in light of this context.

to the operation and then stepped back, on the agreement that he was to be kept fully informed.[47] The position taken by Millionshchikov in sanctioning this action stands perhaps as an indication of a degree of trust operating between these scientists; but this has wider significance since it is highly unlikely that he was acting without authorization from Moscow.

The Executive Committee placed Marcovich in charge of the delegation to Hanoi. Marcovich was a senior figure at the Pasteur Institute in Paris, which had 'sister' institutes around the world, including in former French colonies. An *ordre de mission* from the Institut Pasteur provided the 'cover' for his involvement in what was, in effect, a rather clandestine unofficial diplomatic mission. Since it was deemed "essential that this affair be completely in the hands of Pugwash, no aid was requested from official organizations," rather it was financed from "private circles" within France, arranged by Marcel Bleustein-Blanchet.[48] The most immediate problem lay in finding the means to "to open the door to Hanoi." For this, Marcovich, Bauer and Perrin first approached the journalist and leading figure in the French Resistance, Emanuel d'Astier de la Vigerie, who declined on health grounds. However, at Bauer's suggestion, d'Astier sent a telegraph to Raymond Aubrac, fellow former resistance fighter and now a high-ranking civil servant at the UN Food and Agriculture Organization (FAO) in Rome, whom he knew well, and whom he knew also to have friendly relations with Ho Chi Minh. Aubrac's long-standing links to the People's Republic of Vietnam and Ho Chi Minh now came decisively into play. Previously, Aubrac had twice refused to be involved in initiatives to approach Hanoi: what seemingly appealed to him about Pugwash—with which hitherto he had had no contact—was its "reputation" and that it was a "scientists' organization."[49] Aubrac came to Paris on 26 June and in meetings at Bauer's home spoke with Feld, Perrin and Marcovich after which he agreed to take part in the operation.[50]

Aubrac's involvement was crucial. During the Second World War, he had been a leading member of the French Resistance and, after the liberation, was appointed regional Commissioner of the South of France, based in Marseilles. Forty miles outside Marseilles there was a camp for Vietnamese workers brought to France at the outset of the war and kept there under Vichy and German rule to work in munition factories: conditions were terrible, with many deaths. Aubrac put the French administrators into jail and put in place a system whereby the Vietnamese took over the organization of the camp—earning Aubrac the enduring respect of the Vietnamese. During Ho Chi Minh's visit to France immediately after the war for negotiations about newly-liberated areas in the north Indo-China region, he met Aubrac, who was strongly supportive of the Vietnamese fight for national liberation. Their friendship began at a social gathering, where Ho Chi Minh complained to Aubrac that he did not like the hotel in which he was staying, at which point the Frenchman invited

---

[47] Ibid, 5.

[48] Ibid, 3.

[49] Ibid, 2–3.

[50] Undated/anonymous document: An account of the Pugwash intervention in/about Vietnam in 1967–1968. RTBT 5/3/1/36 (3).

Ho Chi Minh to stay as a guest in his own house where he remained for six weeks. In addition to cementing their friendship, there were occasional visits at Aubrac's house from the Prime Minister, Pham Van Dong. Aubrac and Ho Chi Minh stayed in touch during the 1950s, for example, through Aubrac's role in the French delegation at the Geneva Conference in 1954, where the agenda included the future of Vietnam.

In 1967, the FAO granted Aubrac leave of absence when he explained that this was in order to carry out some work for the Pugwash organization, although Galtung reports he did not disclose the nature of the project. The PCSWA was an important part of the explanatory narrative about the absence of both Aubrac and Marcovich from their respective places of work and was also given as part of the reason for their trip into the war zone. For example, Marcovich travelled with documents "to substantiate the cover of setting up a regional Pugwash Conference." This "cover story" was deemed necessary to protect the true character of the operation. According to Galtung, they arrived in Phnom Penh on 19 July, where the North Vietnamese Embassy was contacted and informed that Aubrac would like to meet with the President (Ho Chi Minh) and Prime Minister (Pham Van Dong). The requisite visas arrived within twenty-four hours, this process being smoothed by the American journalist Wilfred Burchett.[51] Making it known that they were acting under the auspices of Pugwash, Marcovich met with the Director of the Pasteur Institute in Phnom Penh, the Minister of National Education and the North Vietnamese Minister of Health, Dr. Pham Ngoc Thach. Meanwhile, Aubrac met with the (UN) Food and Agriculture Organization representative in Hanoi and, crucially, also with Ho Chi Minh.

On 28 July, Marcovich and Aubrac returned to Paris via Phnom Penh. Beyond the 'soundings' taken from their meetings in Hanoi, they were also carrying with them various materials, including photos and films of the effects of the American bombing. Initial plans arranged by Bauer to discuss their findings at a meeting at the Elysee and also in a further meeting with Pugwash colleagues were abandoned because the tentatively positive nature of their soundings in Hanoi warranted urgent discussion at the highest levels. To this end, they spent the entire day of their return at Marcovich's home in discussion with Henry Kissinger. A report on the trip was transmitted to Washington, via the classified channel of the US Embassy in Paris, and an exchange of messages between Washington and Hanoi followed. Kissinger shuttled between Washington and Paris in the early weeks of August and on returning to Paris mid-month had hoped that Marcovich and Aubrac would make a second visit to Hanoi. However, this was rendered impossible by the destruction on 15 August by American bombing of the main bridge on the Red River—the Bridge Long Bien/Bridge Paul Doumer. After this, trust in Hanoi in the channel established by the Frenchmen evaporated. Kissinger persisted, but the message he carried from Washington was rejected by Marcovich and Aubrac who were not prepared to transmit it to Hanoi

---

[51] Undated document: 'An account of the Pugwash intervention in/about Vietnam in 1967–1968,' 3. RTBT 5/3/1/36 (3). In August 1945, Burchett had been the first American journalist to reach Hiroshima and his report on the devastation he saw there, published in *Harper's Magazine,* sent shockwaves through a large swathe of American society and brought him into conflict with Washington. In some quarters, Burchett came therefore to be regarded as a radical and controversial figure.

because it included terms they knew would be unacceptable to the leadership of the People's Republic of Vietnam. This message was then "referred back to the highest level" in Washington where changes insisted upon by the Frenchmen were accepted and the message revised. On 22 August, Hanoi refused to give visas to transmit the (revised) message—a decision perhaps influenced by the intensified US bombing campaign. After consulting with Rotblat, who was in Paris, the Frenchmen continued in their efforts and, on 25 August, were finally able to pass the message on to Mai Van Bo at the North Vietnam Embassy in the French capital.

However, before the message was received in Hanoi, the US had carried out another round of heavy bombing of the city: the message was now deemed unacceptable by the leadership. Aubrac and Marcovich asked Kissinger to urge Washington to cease bombing Hanoi—which did eventually stop, only for the US to begin bombing Haiphong (Anderson 2002; Kissinger 2003; Lawrence 2010). Between late August and mid-October 1967, Galtung reports a "continuous transfer of messages both ways using Marcovich as a go-between, in constant consultation with Kissinger and Aubrac." Typically, the messages dealt with making the terms more precise. However, American bombing continued, and the Vietnamese eventually broke the channel. In a speech by Lyndon B. Johnson on 19 March 1968 which came to be seen as an important step on the road to peace, Johnson referred to a message given to Hanoi in August 1967—which Johan Galtung reports was the message transmitted by Marcovich and Aubrac via Mai Van Bo. Talks to end the Vietnam War began in Paris in May 1968. Galtung reported that Henry Kissinger later "confirmed" that the channel created by Pugwash—"was the only channel operating during this period."[52]

This channel of communication involved Marcovich and Aubrac acting as couriers carrying messages between Washington and Hanoi—with the consent of, if not on behalf of, Washington. Working informally, under the auspices of Pugwash, there was knowledge of their activities at the highest political levels both in Washington and on the Soviet side, via Millionshchikov. On the American side, Henry Kissinger played a decisive role, acting as a pivot between Robert McNamara of the US State Department (in 1967 working out of the US Embassy in Paris), and Ho Chi Minh, Pham van Dong and Mai van Bo in Hanoi. In effect, the Marcovich-Aubrac mission constituted a form of what Allan Pietrobon has called "citizen diplomacy," in this case initiated by Pugwash scientists and carried out under the auspices of the Pugwash organization (Pietrobon 2016).

This intervention also marked a radical departure for Pugwash in being unconnected to nuclear weapons, in tackling a war prosecuted by the western superpower, and in drawing centrally upon on the wider networks of (French) Pugwash scientists. It took place in a context within Pugwash where divisions over the Vietnam War were playing out in the public space of its conferences and, internally and in private, were threatening to tear the organization apart. This points to a divide between on the one hand the outward-facing world of the conferences and, on the other, the sequestered enclave of the Continuing and Executive Committees. The

---

[52] Johan Galtung. 'The role of Pugwash as a communication link between Washington and Hanoi.' Report dated 12 September 1968. RTBT 5/3/1/22-2, 5.

creation and maintenance of what, in practice, approximated to distinct spheres of operation within Pugwash, was made possible by the combination of its hierarchical structure and its informal *modus operandi*. This accorded a degree of privacy—and crucially autonomy and agency—to the leadership vis-à-vis the wider organization which enabled senior Pugwash scientists from East and West to forge a role in unofficial diplomacy beyond the conferences. The creation by the small 'inner circle' of the Pugwash leadership of a shielded realm of 'discretion' enabled them also to draw a line—albeit contingent and blurry—between the politics of the superpower powerplay and the world they inhabited within the PCSWA. This world resided in the relationships pertaining between these scientists. The understanding—if not always agreement—and goodwill built between these scientists over time meant that, in times of crisis in particular, they were able to suspend the realities of the Cold War world around them long enough to develop the means for a distinctive brand of informal diplomacy unique to the PCSWA.

4.6   Pugwash and Its Scientists: Pioneering Track II Diplomacy?

Taken together, the Cuban, LTBT and Vietnam interventions cast fresh light on how Pugwash scientists formulated 'back-channel' responses tailored to particular situations. On the one hand, this was made possible by changing perceptions of the Pugwash organization within senior government circles in Washington and in London—who, after the conference in Moscow and also in Stowe—came to perceive it as potentially of use to them. Scientists close to the Kremlin and to the White House in conversation around the Pugwash table at these conferences served as a platform for a widening role in informal diplomacy. On the other hand, this reflected the determination, wit and resourcefulness of senior Pugwash scientists on the Continuing and Executive Committees. To a large degree, the Cuban and Vietnam interventions rested on the respect accorded the scientists of Pugwash, both as scientists but also as scientists trying to work for peace, and on the organization's track record in this direction at its conferences. The PCSWA's contribution to the LTBT process rested centrally on the technical and scientific expertise of its scientists, but was also enabled by their connections within government circles (Doel 1997). To be sure, access to these circles could not always and/or straightforwardly be equated with wielding influence within them but, on this occasion, the evidence indicates this to have been the case.

Taking a wider view, these interventions by the PCSWA and its scientists in East and West in the unofficial diplomatic realm rested on their status as natural scientists: they had relevant expertise and were seen as having integrity and being trustworthy. In a pragmatic and contingent process, Pugwash scientists developed their own ways of working in the international realm at the intersection between science, politics and policy-making during the 1960s Cold War. In this period, the limits and/or failures of official diplomacy created the need for alternative forms of communication between East and West. As senior Soviet Pugwash scientist Lev Artsimovitch put it in 1963, official "Diplomacy" was "an antiquated vehicle. Our role is to remove obstacles from the path of this antique chariot" (Rubinson 2011, 291).[53] Pugwash scientists were in the vanguard of forging—both at and beyond the conferences—an alternative to this "chariot" developing a new kind of informal but politically meaningful communication that came in this period to be called Track II diplomacy. The unofficial and informal character of Track II conversations and exchanges by their very nature made for considerable diversity in terms of the emerging modes, mechanisms and range of actors involved in this form of politically sensitive communication. At Pugwash, in the extra-conference setting, Track II initiatives typically took the form of cables, 'phone calls and letters, and small, private meetings between scientists close to both the American and Soviet governments. The leadership learned quickly from experience. All three case studies—Cuba, the LTBT and Vietnam—involved activating and mobilizing the primary resource of the PCSWA, that is to say, its international, cross-bloc network of scientists.

Viewed in the context of what was happening at the time within Pugwash, these unofficial diplomatic interventions enable us to see more clearly the distinction between the PCSWA as a whole, and the leadership. In branching out beyond the conferences, they also illuminate an emerging feature of the internal dynamics within the Pugwash organization, namely the way in which, with decision-making power concentrated in its hands, the Continuing Committee and in particular, from 1963 onwards, the members of the Executive Committee, functioned with a growing degree of autonomy. These Committees—on which the Secretary General had a permanent seat—had enormous discretionary powers. Drawing on the wider resources of the PCSWA, members of these Committees conceived, coordinated and carried out novel modes of informal diplomacy distinctive to Pugwash. Acting under the Pugwash umbrella, the inner circle of leadership was operating, in effect, as an elite within an elite. In the course of these interventions, this elite formed the vanguard of an approach to East–West relations centred on the kinds of encounters, exchanges and activities that came to be called Track II diplomacy.

These interventions also point to the signal importance of Joseph Rotblat. The Secretary General was consistently at the heart of informal diplomatic initiatives mounted by the Pugwash organization. In orchestrating and coordinating these activities he was, in effect, also fashioning and honing his own distinctive brand of "citizen diplomacy" within Pugwash, serving variously as broker and mediator within the Continuing and Executive Committees, and with actors beyond the PCSWA (Pietrobon 2016). Here Rotblat's impartiality as Secretary General was a valuable resource: in practice, it translated into a sign of integrity that earned him a great deal of goodwill from colleagues on both sides of the bloc divide—and beyond.

There was also the need to protect the organization's hard-won role and reputation with state actors as a reliable and discreet forum for informal back-channel diplomacy.

---

[53] Artsimovitch made this comment during private Pugwash meeting at the Ciba Foundation in London in March 1963 where the Black box concept was discussed.

A key strategy included keeping quiet about these activities. They were reported only very briefly and retrospectively in the minutes of the Continuing Committee. In terms of external audiences, we can again see Rotblat's controlling hand: he remained firmly in sole charge of the portrayal of the PCSWA in the wider public realm, publishing various pamphlets and books with a carefully crafted narrative about its work that emphasized the integrity of its scientists and their earnest attempts to work for peace. In these accounts the role of Pugwash scientists in the realm of informal diplomacy was consistently glossed over and downplayed whilst efforts in more 'conventional' areas in which science was bound up with diplomacy, such as initiatives fostering international scientific cooperation, were highlighted. For example, in his 1967 history of the PCSWA, Rotblat included extensive coverage of the first Stowe conference, whilst both the Moscow and second Stowe conferences received comparatively cursory treatment. Neither back-channel successes nor internal tensions/divisions were dwelt upon. In this way, the more politically sensitive activities of the PCSWA were kept out of the spotlight—in order that they might continue.

The deep divisions over Vietnam precipitated a growing sense of the crisis within the Pugwash organization. But there were other sources of tension, not least leadership concerns about the Pugwash Study Group on European Security (PSGES). This had been formed in 1965 in response to calls from both Western and Eastern European scientists for issues of concern to them to be placed higher on the Pugwash agenda—namely, European security, especially the German question. But within two years, the Continuing Committee was increasingly uneasy about the widening scope of the PSGES and its modes of working. Indeed, the PSGES came to be seen as a challenge to both the authority of the Continuing Committee and its jurisdiction over all parts of the international PCSWA. By 1967, wherever the Committee looked, it faced problems that threatened the internal cohesion and stability of the Pugwash organization.

References

Aboltin, Vladimir Y. 1966. Main problems of progress towards GCD. In *Proceedings of the 16th Pugwash Conference, Sopot, September 1966*: 111–119.

Anderson, David L. 2002. *The Columbia history of the Vietnam War*. New York: Columbia University Press.

Barth, Kai-Henrik. 2006. Catalysts of change: Scientists as transnational arms control advocates in the 1980s. *Osiris* 21 (1): 182–206.

Barth, Kai-Henrik. 1998. Science and politics in early nuclear test ban negotiations. *Physics Today* 51 (3): 34.

Bridger, Sarah. 2015. *Scientists at War. The ethics of Cold War weapons research*. Cambridge, MA: Harvard University Press.

Doel, Ronald E. 1997. Scientists as policy-makers, advisors and intelligence agents: Linking contemporary diplomatic history with the history of contemporary science. In *Historiography of contemporary science and technology*, ed. Thomas Soderqvist, 215–244. Amsterdam: Harwood Academic.

Dubinin, M. 1966. Vietnam—the scene of employment of mass destruction weapons. In: *Proceedings of the 16th Pugwash Conference, Sopot, September 1966*: 131–136.

Emelyanov, Vasily. 1966. Pugwash Conference. *Moscow News*, October 8.

Feld, Bernard T. 1966. General remarks on current problems of disarmament and European security. In *Proceedings of the 16th Pugwash Conference, Sopot, September 1966*: 71–75.

Kadomtsev, Boris B. 1985. *Reminiscences about Academician Lev Artsimovitch*. The Soviet Academy of Sciences: Nauka.

Kissinger, Henry. 2003. *Ending the Vietnam War: A history of America's involvement in and extrication from the Vietnam War*. New York: Simon and Schuster.

Kistiakowsky, George B. 1974. Presidential science advising. *Science* 5 April 1974 (184): 38–42.

Kraft, Alison. 2018. Dissenting scientists in early Cold War Britain. The "Fallout" controversy and the Origins of Pugwash, 1954–1957. *Journal of Cold War Studies* 20 (1): 58–100.

Kraft, Alison. 2020. Confronting the German Problem: Pugwash in West and East Germany, 1957–1964. In *Science, (anti)communism and diplomacy: The Pugwash Conferences on Science and World affairs in the early Cold War*, eds. Alison Kraft, and Carola Sachse, 286–323. Leiden: Brill.

Kubbig, Bernd W. 1996. *Communicators in the Cold War: The Pugwash Conferences, The U.S.-Soviet Study Group and the ABM treaty. Natural scientists as political actors: Historical successes and lessons for the future*. PRIF Reports No. 44. Frankfurt am Main: PRIF.

Lawrence, Mark Atwood. 2010. *The Vietnam War: A concise international history*. Oxford: Oxford University Press.

Lüscher, Fabian. 2020. Party, peers, publicity. Overlapping loyalties in early Soviet Pugwash, 1955–1960. In *Science, (anti-)communism and diplomacy: The Pugwash Conferences on Science and World Affairs in the early Cold War*, eds. Alison Kraft, and Carola Sachse, 121–155. Leiden: Brill.

Millionshchikov, Mikhail D. 1965. Disarmament and international cooperation: A way of struggle against the threat of war. In *Proceedings of the 14th Pugwash Conference, Venice, April 1965*: 168–176.

Millionshchikov, Mikhail D. 1966. What is needed for ensuring European security. In *Proceedings of the 16th Pugwash Conference, Sopot, September 1966*: 58–70.

Olšáková, Doubravka. 2020. Czechoslovak ambitions and Soviet politics in Eastern Europe: Pugwash and the Soviet peace agenda in the 1950s and 1960s. In *Science, (anti-)communism and diplomacy: The Pugwash Conferences on Science and World Affairs in the early Cold War*, eds. Alison Kraft, and Carola Sachse, 259–285. Leiden: Brill.

Pietrobon, Allen. 2016. The role of Norman Cousins and Track II diplomacy in the breakthrough to the 1963 LTBT. *Journal of Cold War Studies* 18 (1): 60–79.

Rabinowitch, Eugene. 1963. Pugwash-COSWA: International conversations. *Bulletin of the Atomic Scientists* 19 (6): 7–12.

Rotblat, Joseph. 1965. Address of the Secretary General. In *Proceedings of the 14th Pugwash Conference, Venice, April 1965*: 72–74.

Rotblat, Joseph. 1967. *Pugwash—A history of the Conferences on Science and World Affairs*. Prague: Czechoslovak Academy of Sciences.

Rubinson, Paul. 2011. "Crucified on a cross of atoms." Scientists, politics and the Test Ban Treaty. *Diplomatic History* 35 (2): 283–319.

Rubinson, Paul. 2020. American scientists in "Communist Conclaves." Pugwash and anti-communism in the US. In *Science, (anti-)communism and diplomacy: The Pugwash Conferences on Science and World Affairs in the early Cold War*, eds. Alison Kraft, and Carola Sachse, 156–189. Leiden: Brill.

Salomon, Jean-Jacques. 2001. Scientists and international relations: A European perspective. *Technology in Society* 23: 291–315.

Wolstenholme, Gordon E.W. 1949. The Ciba Foundation. *Nature* 164 (4159): 118–119.

Chapter 5
The Politics of European Security: A Step Too Far for Pugwash?

Abstract This chapter is in two parts. First, it charts the growing influence of Europeans within the Pugwash organization during the 1960s and their efforts to put European security onto its agenda. This culminated in the creation in 1965 of the Pugwash Study Group on European Security (PSGES). The PSGES was an experiment: focused on problems in international politics without reference to science and technology, it brought into Pugwash scholars from law, economics and the social sciences. This departed from Pugwash tradition: it posed enormous challenges for the Continuing Committee and tested both the limits of the organization's political engagement and its capacity for interdisciplinarity. The second part of the chapter explores the deepening crisis engulfing the Pugwash leadership by 1967 as it grappled with the PSGES problem, faced internal divisions over Vietnam, lost its monopoly as an international forum for informal dialogue about disarmament, and struggled to meet the emerging challenges of the mid-1960s Cold War.

Keywords Pugwash Study Group on European Security (PSGES) · The German problem · Ivan Málek · Joseph Rotblat · František Šorm · Antonin Šnejdárek · Karlovy Vary conference · Ronneby conference · Pugwash Symposia

In the mid-1960s, the study group approach became important as a strategy to enable the PCSWA to engage with new problems emerging as a result of the changing dynamics of the Cold War. Study Groups appealed already in the late 1950s to the Pugwash leadership as a means both to mitigate the limitations inherent in an organization that met once a year and to enable sustained, in-depth work on a specific topic. The idea for an East–West study group on disarmament conceived at the Moscow conference in 1960 had been taken forward—albeit outside the PCSWA, namely, the SADS group. Some national Pugwash groups, notably that in West Germany, readily embraced the study group approach as a means to engage with problems important within the national context (Albrecht et al. 2009, 366–369). In 1962 in his report as Secretary General at the tenth conference in London, Joseph Rotblat shared his view that in the future Pugwash would have to "rely heavily" on such groups.[1] This soon became reality with the creation in 1964 of a group on Biological Weapons (BW) and in 1965 of the Pugwash Study Group on European Security (PSGES).[2] Significantly, both initiatives originated in circles outside the Continuing Committee.

The creation of these international and cross-bloc groups, which met outside the conference framework, signaled the widening scope of the international PCSWA. But they pointed to another important shift taking place within the organization: its agenda was no longer being set solely by the Continuing Committee, that is to say by physicists mainly from the US, the Soviet Union and the UK. Of course, the formation of the PSGES and the BW groups had first to be authorized by the Committee, which retained its decision-making power over the development of all parts of the international Pugwash organization. Nevertheless, a process was underway in which new constituencies—in the case of the PSGES, scientists from Western and Eastern Europe and Scandinavia, and in the case of the BW group, biologists—were driving the PCSWA in new directions.

The thirteenth conference, held in Karlovy Vary in 1964, was decisive in the formation of both study groups. This conference, along with that in Dubrovnik a year earlier, marked a watershed in terms of European influence within the PCSWA (Kraft 2020). The PSGES was one outcome of this influence, with decisive steps taken within Working Group 1 in Karlovy Vary. Meanwhile, the move to create a study group on biological weapons at Karlovy Vary was driven by the WHO-based American microbiologist Martin M. Kaplan working closely with colleagues from Czechoslovakia, Denmark, Sweden, and the UK.[3] Kaplan had a long-standing interest in the issue of biological warfare: in 1959, he had been instrumental in making both biological and chemical weapons the theme of the fifth PCSWA held in Pugwash, Nova Scotia (Rotblat 1967, 24–26, 100–104; Perry Robinson 1998). At the time, this broke new ground in being the first international conference devoted specifically to these distinctive kinds of weapon, which in 1947–1948 had been placed by the UN into a separate category. Resurgent concerns in the 1950s about biological weapons reflected suspicions about American use of them during the Korean War (Kaplan 1999).

The two new study groups followed very different trajectories within the PCSWA. Whereas the BW group came widely to be regarded as effective and successful, the PSGES rapidly became a source of great concern to the Secretary General and some of his colleagues on the Continuing Committee. This chapter focuses in particular on the PSGES, charting its origins and development, analyzing when and why it came

---

[1] Joseph Rotblat. Memo on future activities and organization. London, 1962. RTBT 5/2/1/10.

[2] Continuing Committee meeting no. 20, December 1964, Ciba Foundation, London. RTBT 5/3/1/2 (Part I).

[3] Martin Kaplan, Carl Göran Heden, John Humphrey, Ole Maaloe and Ivan Málek. Proposals for a Study Group on Inspection for Biological Warfare Weapons as a Pilot Scheme in Central Europe. Paper at PCSWA 13, Karlovy Vary, September 1964, XIII-10, RTBT 5/2/1/13 (2).

to be seen as a problem by the Pugwash leadership, and the process by which the Committee dissolved the group in 1968. Beyond its value as a means to explore the strengthening European dimensions of the PCSWA in the mid-1960s, the PSGES also provides a window onto the politics and power relations at work within the organization.

5.1 Pressure to Change the Agenda: European Scientists, European Problems

From the outset, the international PCSWA aspired to having a global reach. Yet, the early dominance of the superpowers and the UK within Pugwash meant that the organization carried within it an asymmetry that, left unchecked, could create a potentially damaging core/periphery dynamic between the leadership and continental Europeans from both sides of the bloc divide. Some sense of such frustrations was occasionally discernible. For example, in a report on the Cambridge and London conferences in 1962, the West German Pugwash group noted that scientists from smaller European countries stood 'on the margin' ("am Rand") of discussions wholly dominated by the US, USSR and UK.[4]

The creation in 1959 of the European Pugwash Group (EPG) provided, as the name suggests, a forum for Europeans to come together and discuss issues of concern to them (Kraft 2020). Funded privately, largely by the wealthy American James Wise, the organization and management of the EPG was overseen by Martin Kaplan, working closely with Joseph Rotblat: both attended all its meetings.[5] Wise and Kaplan had private homes in Geneva which served as the venue for the EPG, which met regularly on an informal basis outside of the annual conferences every six months between spring 1959 and autumn 1961, and less regularly until 1964.[6] Meetings typically involved between ten and fifteen scientists from across Europe—initially, from western Europe, with colleagues from Yugoslavia and from within Eastern Europe involved from 1961 and 1963 respectively.[7] Those involved included Hans

---

[4] Bericht über die 4. Mitgliederversammlung der VDW e.V. am 27–28.10.1962 in Marburg-Lahn. RTBT 5/2/1/11 (29). ‚Die Diskussionen waren beherrscht durch Vertreter der USA und der UdSSR mit Einschaltung Großbritanniens; die übrigen kleinen Nationen standen am Rand,' 3.

[5] An author and educator who in the 1930s had warned against the Nazis, James Wise was also the editor of *Opinion,* a journal of Jewish Arts and Letters. Wise lent substantial financial support to the PCSWA with the stipulation that this be kept private. For example, in 1967 when Rotblat was writing his book on the history of the first decade of Pugwash, Wise asked that no reference be made to his involvement with the organization. Joseph Rotblat to Martin Kaplan, 22nd September 1961. RTBT 5/2/3/5. James Wise, Obituary, *The New York Times,* 30 November 1983.

[6] Minutes of EPG meetings, 1959–1964. RTBT 5/2/3/1-6.

[7] Initially, the scientists involved in the EPG came from: Austria, Denmark, France, Italy, the Netherlands, Norway, Switzerland and West Germany. RTBT 5/2/3/4. Yugoslavian Ivan Supek attended the 4th meeting in April 1961, and participants at the 6th meeting in March 1963 included

Thirring, Hans Tolhoek, Karol Lapter, and Herbert Marcovich (from Austria, the Netherlands, Poland and France, respectively), the West German Gerd Burkhardt and, later, the East Germans Günter Rienäcker and Max Steenbeck. Initially geared to supporting the formation of national groups, the EPG soon came to serve as a place for airing and sharing views on the development of Pugwash and on its conference program.[8] The EPG functioned as a European hub outside of the conferences where trust was built between European scientists along with a sense of pan-European solidarity. In the early 1960s, this cross bloc European network began to advance their interests within the Pugwash organization. Eyeing the formation of SADS and moves to create a Biological Weapons group, the EPG also began to press for a study group dedicated to European security.[9]

Strengthening European involvement in Pugwash was given impetus by profound shifts in the wider geopolitical landscape, most obviously the Berlin crisis. The volatile stand-off between Bonn and East Berlin was creating deep alarm within the countries of Central Europe. Within Pugwash, concerns about the intractable German-German conflict served as a rallying call to Europeans from East and West alike, lending momentum and focus to the EPG (Kraft 2020). Poland and Czechoslovakia in particular, as near neighbours, saw themselves as most affected by the German question: for František Šorm, the 'burning question' about Germany constituted the primary 'source of danger' in Europe.[10] Indeed, Eastern Europeans played an increasingly prominent role in discussions and activities relating to the political situation in Central Europe: as Rotblat later noted, 'much greater interest was taken in European problems by the socialist bloc countries.'[11]

In 1962, senior figures within the EPG—the Polish historian Karol Lapter, and the Dutch and Austrian physicists Hans Tolhoek and Hans Thirring—picking up on some ideas floated in the Rapacki Plan, raised with Rotblat their idea for a meeting in Geneva on the topic of 'disengagement' in Central Europe. In February 1963, the Continuing Committee approved the plan and issued invitations to the meeting in Geneva in early March 1963.[12] The *Disengagement in Europe* meeting, which was the 6th gathering of the EPG, took place in Geneva between 2 and 4 March 1963 and involved sixteen scientists from twelve countries.[13] Participants were actively

---

the East German physicist Max Steenbeck, Ivan Málek and Theodor Němec from Czechoslovakia, and Karol Lapter.

[8] Pugwash European Group, Meeting No 1, March 1959. RTBT 5/2/3/1.

[9] The idea for the study group was included on the proposed program for EPG meeting in Geneva, 2–4 March 1963. RTBT 5/4/12/5 (1). See also: Minutes of EPG meetings, 1959–1964. RTBT 5/2/3/1-6.

[10] Frantisek Šorm. 'Remarks on past and future activities,' 09/66, Paper XVI-18. 16th Pugwash Conference, Sopot, Poland. RTBT 5/2/1/16 (8).

[11] Joseph Rotblat. 1975. The Pugwash Movement in Europe. RTBT G75.

[12] The records of the EPG are for some meetings incomplete and in places in some disarray. RTBT 5/2/4/1-8.

[13] 'Notes on meeting of European representatives on Disengagement in Europe,' Geneva, 2–4 March 1963. RTBT 5/2/17/24. Meetings of the Continuing Committee: No. 16, 8–10 February 1963, London and No. 17, Dubrovnik, September 1963. RTBT 5/3/1/2 (Pt 1) (4). In addition to Lapter,

discouraged from publicizing the meeting, and were also strongly reminded that it took place under Chatham House rules. The German problem was high on the agenda, as were the on-going difficulties surrounding the participation of East German scientists at Pugwash conferences (Kraft 2020).[14] A pre-circulated paper by Lapter and Tolhoek—an East–West collaboration—entitled 'General principles for a zone of disarmament in Europe,' provided the starting point for discussion which soon turned to the need for Pugwash to engage much more strongly with the German question.[15]

Dialogue between West and East German scientists was seen as a prerequisite for tackling the German problem. The EPG was an early site where this began. In Geneva, West German Gerd Burkhardt and East German Max Steenbeck produced a five-point document which was a starting point for a co-written paper on the German question: the plan was to complete this in time for the upcoming eleventh PCSWA in Dubrovnik. Rotblat would later recall that this provided a powerful spur to engagement within Pugwash with the German problem (Rotblat 1967, 45).[16] The sixth EPG meeting put German-German relations within Pugwash on a new footing and also led to measures that facilitated regular East German participation at the Pugwash conferences, beginning in Dubrovnik in September 1963. All of these developments signaled the growing confidence of Europeans within the organization.

Growing European influence was strikingly in evidence at the 11th Pugwash conference in Dubrovnik which was sponsored by the Yugoslav Academies of Science and organized by Ivan Supek. It was given to the theme 'Current problems of disarmament and world security,' and was attended by seventy-eight delegates (including fourteen observers) from twenty-four countries. Meeting one month after the LTBT had been signed in Moscow, hopes were high of a shift in the superpower relationship that might lead to further progress towards disarmament. In this climate, Europeans redoubled their efforts to reposition European security higher on the agenda of the international PCSWA. Certainly, a new mood was in evidence in Dubrovnik where, for example, the participation of scientists from both German

---

Tolhoek and Thirring, and the Germans, those present were: Málek and Němec (Czechoslovakia); Valkenburgh (the Netherlands); Aubert (Norway); Houtermans (Switzerland); Rotblat and Lindop (UK); Kaplan (US); Jaksic and Supek (Yugoslavia).

[14] 'Notes on Disengagement meeting,' RTBT 5/2/17/24, 3. Seventeenth meeting of the Continuing Committee, September 1963, Dubrovnik, 2. RTBT 5/3/1/2 (Pt 1). East German scientists wishing to attend Pugwash conferences held in NATO countries faced difficulties on account of the Hallstein Doctrine pursued by Bonn.

[15] The minutes of the sixteenth meeting of the Continuing Committee, 8–10 February 1963, Ciba Foundation, London, record the Geneva meeting as involving discussions about creating a Study Group on the theme of *Disengagement*/an 'atom free zone' in Europe. RTBT 5/3/1/2 (Pt 1) (4). 'Notes on meeting of European representatives on Disengagement in Europe,' Geneva, 2–4 March 1963. RTBT 5/2/17/24, 8 and Appendix 1, 6.

[16] Notes on meeting of European representatives on Disengagement in Europe. Geneva, 2–4 March 1963. RTBT 5/2/17/24, Appendix 3, The German problem. Gerd Burkhardt/Max Steenbeck, 'The German problem and its relevance to regional and limited disarmament agreements in Central Europe.' Dubrovnik, September 1963. Main papers: XI.12-Burkhardt/XI.13-Steenbeck. RTBT 5/2/1/11 (11).

states created new possibilities for informal dialogue between them, including on the German question, especially within Working Groups.

Working Group 3 was given to the topic of 'Denuclearized Zones, especially in Central Europe and the Balkans,' and took its cue from Tolhoek and Lapter's paper at the EPG 'Disengagement' meeting in March.[17] Table 5.1 summarizes membership of this group which included a mix of East and West Europeans, including from both Germanies, and Henry Kissinger (Table 5.1). Tackling this theme was a major step that took the PCSWA into new and sensitive political terrain directly related to the policies of both superpowers in Central Europe. The concept of denuclearized zones in this region was controversial because it was fundamentally bound up with the German question. Certainly, this idea was anathema to West German Chancellor Konrad Adenauer because it went against Bonn's conceptions of NATO and of the country's role in the defence of Europe (Schrafstetter 2004). Discussions in Working Group 3 ranged across the definition and geographical parameters of denuclearized zones, the staging of their introduction in Central Europe, and the value of such zones as a brake on nuclear proliferation and as a step towards General and Complete Disarmament.

By convention, each Working Group produced a summary report to be circulated and discussed in plenary session. These reports included findings arising from the conversations between members of each group: in effect, these discussions constituted a novel form of specialized technopolitical research. The final report of Working Group 3 asserted that the creation of a denuclearized zone in Central Europe (defined as comprising Czechoslovakia, Poland and both Germanies) could "help the East

Table 5.1 Working Group 3, Dubrovnik, September 1963[18]

Topic: Denuclearized zones, especially in Central Europe and the Balkans	
Convenors: Karol Lapter (POL) and Hans Tolhoek (NE)	
H. Afheldt (FRG)	H. Kissinger (US)
R. Bognar (HUN)	V. Knapp (CZE)
N. N. Bogoliubov (USSR)	G. Nadjakov (BUL)
G. Burkhardt (FRG)	G. Rienäcker (GDR)
A. A. Buzatti-Traverso (IT)	J. Rotblat (UK)
B. Dimissianos (GRE)	I. Supek (YUG)
D. Dumitrescu (ROM)	L. Szilard (US)
B. Glass (US)	P. Valkenburgh (NE)
L. Infeld (POL)	

[17] In Dubrovnik, there were five Working Groups. The other four were: 1. Problems of general disarmament. 2. Consequences of the spread of nuclear weapons. 4. Role of non-aligned nations in disarmament and world security. 5. The Partial test-ban, the problems of detection, and the next steps. Sixteenth meeting of the Continuing Committee, February 1963. RTBT 5/3/1/2 (Pt 1) (4).

[18] *Proceedings of the 11th Pugwash Conference, Dubrovnik, September 1963.*

and West German governments to make a real effort to diminish the existing tension between them" that "may lead to removal of the obstacles to genuine communication (including travel between their territories)."[19] Elsewhere, there were criticisms of the status quo in Central Europe, of US policy in the region and the West German government, especially Bonn's reluctance to enter into discussions about a denuclearized zone in the region. The report also urged Pugwash to make efforts in this direction:

> We are of the opinion that it will be most useful for the 11th PCSWA to appeal to all governments directly concerned with the situation in Central Europe, and to urge them to enter into negotiations leading to the lessening of tensions in this area and to the establishment of a denuclearized Central Europe. Thus we may hope to achieve a peaceful Central Europe and bring nearer the ultimate unification of Germany.[20]

This Report meant a great deal in particular to the East German Pugwash group. Its scientists repeatedly called for its recommendations to be upheld and for the PCSWA to publicly endorse them, and it later defined the GDR position in any/all discussions within the PSGES.

Such forceful and partisan statements from Working Groups ran counter to the quiet diplomacy—not to mention the cherished principle of political neutrality—of Pugwash (Rotblat 1967, 157–163). Hitherto, the leadership had exercised sole and close control over accounts of the conferences. But the advent of Working Groups which were empowered to write reports on their own terms was beginning to challenge that tradition. Perhaps not by accident, the convenors of these typically cross bloc groups were often seasoned veterans of Pugwash mainly from the US, USSR and UK, including past and present members of the Continuing Committee.[21] But it was not possible for the convenors to control the membership of a group, the discussions taking place within it, or the content of the report arising from it.

Within the Continuing Committee, ripples of worry were spreading about strongly worded statements issued from Working Groups. Although not wholly in the public domain, these reports could—at the discretion of the Continuing Committee—be included in the Conference *Proceedings,* the distribution list of which included national governments. There was therefore a danger that trenchant criticisms of the western alliance, such as those made in Working Group 3 at Dubrovnik, could reach the ear of Washington and London. Indeed, whilst Working Groups had rapidly become an indispensable asset, not least as a major site of work carried out during the conferences, it is clear that the Continuing Committee handled the reports arising from them with care. Sometimes, when the content was potentially politically controversial, these reports were omitted from the conference *Proceedings.* On occasion, as was the case at the eighth conference in Stowe in 1961, two copies of the *Proceedings* were created, with and without the reports of the Working Groups.

---

[19] Denuclearized zones, especially in Central Europe and the Balkans. Report of Working Group 3, Dubrovnik, 09/1963. RTBT 5/2/4/3 (3). [Records of the PSGES, Report No. 20].

[20] This report also included a similar resolution calling for the creation of denuclearized zones in the Balkans, Africa and Latin America.

[21] Conference *Proceedings* include the reports of Working Groups and details on membership.

An emerging pattern in which Working Groups served lobbying purposes, and were adding a new potentially partisan edge to the PCSWA was even more strikingly apparent at the 13th conference held in Karlovy Vary from 13 to 19 September 1964. Hosted by the Czechoslovakian Pugwash Group, it was given to the theme, 'Disarmament and peaceful collaboration among nations' (Maddrell 2005, 624).[22] This marked both the realization of Czech ambitions to host a Pugwash conference and a high point for the Czechoslovak group, which was increasingly active in the PCSWA: Czech microbiologist Ivan Málek (1909–1994) was subsequently appointed to the Continuing Committee (Bull 2011; Olšáková 2018, 2020).[23] Fifty-eight scholars from nineteen countries attended, including many East and West Europeans and the Americans Henry Kissinger and Jerome Wiesner (Rotblat 1967). Karlovy Vary demonstrated further the steadily strengthening sense of cross-bloc European solidarity apparent in the EPG and in Dubrovnik. This was strikingly apparent in the main program—consisting of some thirty-six papers—including several given to various aspects of the German question, for example, by Leopold Infeld on 'The Berlin problem,' by Tolhoek and Šorm on the reduction of tensions in Central Europe, and by the American Hans J. Morgenthau on 'The real causes of tension: Berlin and Central Europe.'[24]

The conference included five Working Groups: 1. Measures for reducing tensions/dangers of war, especially in Central Europe; 2. Current status of proposals for arms limitations; 3. Progress towards comprehensive disarmament; 4. Problems of collective security and 5. Aims and methods for peaceful collaboration among nations. The convenors were appointed by the Continuing Committee and were mainly natural scientists, but membership of the groups was increasingly interdisciplinary, including a mix of science and policy advisors, and social scientists with, for example, expertise in economics, psychology, and international relations. By convention, scientists could participate in up to two groups, for example, the West Germans Gerd Burkhardt and Horst Afheldt and the East German Günther Rienäcker participated in groups 1 and 4; meanwhile Rienäcker's countrymen were also in two groups, Peter Hess was in groups 1 and 5, while Jürgen Kuczynski worked in groups 3 and 5 (Green 2017).[25] Also in group 1 were Šorm, Tolhoek, Infeld and Henry Kissinger: Kissinger was also involved in group 4, along with fellow American Jerome Wiesner. (Table 5.2). Notably, there was a considerable degree of overlap in the membership

---

[22] This was attended by the West Germans Burkhardt and Horst Afheldt, and from the GDR, the trusted Rienäcker, Peter Hess and the economic historian Jurgen Kuczynski—whose involvement was of particular concern in Bonn. Minutes of the twentieth meeting of the Continuing Committee, 19–20.12.1964. RTBT 5/3/1/5. During the Second World War, Kuczynski had been the leader of the German Communist Party in London and head of its underground network.

[23] Minutes of the twentieth meeting of the Continuing Committee, Ciba Foundation, London, 19–20 December 1964. RTBT 5/3/1/2 (Pt 1) (4).

[24] List of papers on the main program, Karlovy Vary, RTBT 5/2/1/13 (2). In a memorandum to the Continuing Committee at Karlovy Vary, Šorm urged it to support the formation of a study group on the German question, which he emphasized, was "an issue of vital interest for our people." Frantisek Šorm. On the question of Study Groups. RTBT 5/2/1/13 (43).

[25] Karlovy Vary, List of Working Groups. RTBT 5/2/1/13 (3).

Table 5.2 Working Group 1, Karlovy Vary 1964

Theme: Measures for reducing tensions and the dangers of war, especially in Central Europe	
Convenors: Bentley Glass (US), Jules Moch (FRA) and N. A. Talensky (USSR)	
H. Afheldt (FRG)	H. Kissinger (US)
P. V. Andreyev (USSR)	J. Maddox (UK)
N. N. Bogoliubov	H. Morgenthau (US)
L. R. B. Elton (UK)	G. Rienäcker (GDR)
K. Fowler (AUSTRALIA)	A. Šnejdárek (CZE)
P. Hess (GDR)	F. Šorm (CZE)
L. Infeld (POL)	H. A. Tolhoek (NE)

of Working Group 1 in Karlovy Vary and Working Group 3 in Dubrovnik—a pattern that suggests an on-going conversation between some scientists within Pugwash about European security.

Significantly, in Working Group 1 in Karlovy Vary, East and West Germans were working closely with each other and with colleagues from both sides of the bloc divide. Given its organising theme, discussions here cannot but have touched upon issues of acute concern to their respective governments in East Berlin and Bonn.

In its report, Working Group 1 called for the creation of a Pugwash study group dedicated to European security. This report also levelled fresh and forceful criticisms at the western powers and their policies in Central Europe and began with what, from a Western point of view, was the politically charged recommendation that:

> We consider it urgently necessary that those nations concerned with the German Problem which have not already done so, and in particular the former occupying powers together with the Federal Republic, should recognize and guarantee the existing frontiers of Germany with neighboring states.[26]

Elsewhere, the report also called for the abandonment of the Multi-Lateral Force— a red flag in Bonn—and in a section entitled "Collective Security," it advocated the extension of the UN peace-keeping function and called for a ban on the use of force by any nation in violation of the territorial integrity of another nation. Although not explicit, this recommendation was unmistakeably made with an eye to the situation in Vietnam. In this report, the Pugwash mantra of political neutrality had seemingly evaporated.

Moreover, by now, it had become practice to include summaries of Working Groups in the end-of-conference statement. At Karlovy Vary, the inclusion in the statement of some of the more politically contentious points in the report of Working Group 1 caused great anger in Washington and charges that Pugwash was a

---

[26] Report of Working Group 1: Measures for reducing tensions and the dangers of war, especially in Central Europe. Karlovy Vary, September 1963. RTBT 5/2/1/13 (3).

platform for Soviet propaganda (Rubinson 2020, 172).[27] The Continuing Committee had wide discretionary powers over what was included—and excluded—from conference statements. This raises intriguing questions about why the politically incendiary aspects of this Working Group report were incorporated into the Karlovy Vary statement and why, at this juncture, Pugwash veered into controversial terrain on the international and public stage.

Dubrovnik and Karlovy Vary revealed with new force the paradox of the Working Groups. On the one hand, they could serve to generate fresh thinking and a means for bringing new topics onto the Pugwash agenda. On the other hand, they could be sources of views and proposals that went against the grain of thinking within the Continuing Committee: that is to say, working groups could function as seedbeds of dissent within the PCSWA. The politicized and partisan nature of the reports issued from Working Groups 3 (Dubrovnik) and 1 (Karlovy Vary), together with the conference statement from Karlovy Vary, had implications for relations between the PCSWA and the Johnson administration.[28] At the same time, these developments seeded difficulties for the Continuing Committee in terms of relations with European colleagues who saw the problems of Central Europe as falling within the purview of Pugwash and who were less mindful of the mantra of political neutrality.

The Karlovy Vary and Dubrovnik conferences marked a watershed for the PCSWA in a number of ways. First, the statements from working groups 3 and 1 respectively, parts of which also featured in the conference statements, broke with the Pugwash tradition of steering clear of political controversy. Second, Dubrovnik and especially Karlovy Vary charted a new direction for Pugwash, with the course set by Europeans united by a determination to put European security on the agenda. Third, they placed a Study Group dedicated to the German problem and European security firmly on the agenda.[29] The subsequent creation of this group in December 1965, whilst satisfying the demands of Europeans within Pugwash, also provided the leadership a strategic means to compartmentalize discussion of sensitive issues directly related to American foreign policy in Europe and, moreover, a means to contain discussion of these issues at the annual conferences.

5.2   The PSGES, 1964–1968: A Bold Experiment?

Shortly after Karlovy Vary, Theodor Němec contacted Horst Afheldt and Henry Kissinger with a view to establishing a Study Group dedicated to European security. In December 1964, Němec relayed to Rotblat the news that the Czechoslovakian Pugwash circle—himself, along with Šorm and Antonin Šnejdárek, had taken

---

[27] Statement by the Pugwash Continuing Committee, Karlovy Vary, 20 September 1964. RTBT 5/2/1/13 (1).

[28] For an internal perspective on difficult relations with the US, see: Cecil F. Powell to Joseph Rotblat, 22 August 1966. RTBT 5/2/17/4 (Pt 3).

[29] Karlovy Vary, Report of Working Group 1, 3. RTBT 5/2/1/13 (3).

steps in this direction, including seeking US and Soviet involvement, with Khvostov being the Soviet contact.[30] Plans for the group were modelled on the SADS group. Němec reported that Kissinger had obtained funds for supporting the project, and the Czechs were ready to host the first meeting—indeed, the Czechoslovakian Academy of Sciences issued a standing invitation to the group. Meanwhile, Scandinavians were also mobilizing along the same lines: the Danish chemist (David) Jens Adler had approached Rotblat with a similar proposal. Adler's priorities were to "break down the present impasse in the relationship between the two Germanies," and for the "smaller nations situated in Europe" to tackle the political problems of Europe "intellectually," having a special relationship to the situation "being the first to suffer from tension and conflict, both during an uneasy peace and in war." The Danish and Czech proposals shared a determination for the PCSWA to engage more strongly with the political issues created within the European region by the superpower rivalry. As Adler put it:

> One could perhaps say, that the arms race between east and west has been one of the fundamental causes of present tensions in the central European region. But even though—today—the two main powers in the world, the Soviet Union and the US, wish to improve their relationship and eliminate causes of conflict also in Central Europe, the tension here, induced in a previous period, has got its own basis of existence. The political crisis in Central Europe is growing more and more autonomous.[31]

The PSGES was, then, one response to the changing dynamics of the superpower relationship which were reverberating within and between their respective alliance systems to destabilize and reshape the political constellation of Central Europe (Mastny 2008; Schrafstetter 2004; Trachtenberg 2010). The group was a new departure for the PCSWA in being a European initiative conceived outside of the Continuing Committee. From the outset, it broke new ground in other ways, not least in moving away from an immediate concern with science and technology-related issues. Linked to this, the group was also highly experimental in bringing into the Pugwash fold scholars from disciplines beyond the natural sciences, most prominently from the social sciences, economic and legal disciplines—that is to say, the kinds of expertise and know-how needed for tackling European security. Those taking part in the study group were nominated by the respective national Pugwash groups—although the PSGES co-chairs agreed to submit the attendance lists of its meetings to the Continuing Committee.

Rotblat updated the Continuing Committee on the Czech and Danish proposals which, according to the official minutes, was "pleased" at the prospect of a new Study Group and recommended that Němec and Adler co-ordinate their efforts; he emphasized that representatives from both Germanies be invited to join the project.[32]

---

[30] Němec to Rotblat, 12 December 1964. RTBT 5/2/1/13 (43).

[31] David Jens Adler. Proposal for the creation of a five-nation study group on disarmament in Central Europe. 9 December 1964. Document to be put to the Continuing Committee at its upcoming meeting later that month. RTBT 5/2/1/13 (43).

[32] Minutes of the twentieth meeting of the Continuing Committee, Ciba Foundation, London, December 1964, 8. RTBT 5/3/1/2 (Pt. 1). Both the USA and the USSR were invited to send observers to the group.

In London in early December 1964, Adler took the opportunity to discuss plans for the group with Lapter, Jerzy Sawicki, V. Sojak and the West German, Ekkehart Stein—variously lawyers, historians and/or experts in international relations—when they were all at the joint meeting of the Conferences on Research on International Peace and Security (CORIOPAS)-International Peace Research Association (IPRA).[33] Adler worked closely with Němec throughout 1965 to realize their vision, culminating in the first'Constituting meeting' of the PSGES held between 3 and 5 December 1965 at the Academy of Sciences in Prague.[34] The meeting, which involved five sessions, was attended by Pugwash representatives from Czechoslovakia, Denmark, the Netherlands, Poland and Sweden. Academician Frantisek Šorm gave the opening address.[35] In effect, the PSGES began as an interdisciplinary collaboration between scientists from Eastern and Western Europe and Scandinavia. The report on the Prague meeting took the form of an Aide Memoire—this early use of a term used specifically within the realm of official diplomacy provided an early indication of the culture within the group. Its opening paragraph noted that twenty years after the conclusion of the Second World War, major international problems in Europe—and especially some of those related to the German situation—remained unresolved and that this "constitutes a threat to world security in general and the security of Europe in particular."[36] It went on to set out the focus and priorities of the group, which centered on three principal themes, namely: the German question, European Security and Integrational processes in Europe. With this agenda, the PSGES was taking Pugwash much further into the terrain of international relations and international law.

Following the inaugural meeting in December 1965, the PSGES met seven times between February 1966 and May 1968 with the involvement of scholars from twelve European countries.[37] (Table 5.3). The meetings were funded from private sources, for example, the second meeting in Hälsingborg in Sweden was financed by the Carlsberg company, and/or by national Academies of Science in the Eastern bloc—that of Czechoslovakia being especially generous. Many participants were regular attenders, forming a small cohort that knew each other well: the group was and remained an East–West European project. The Americans were notable by their absence, whilst

---

[33] Adler to Rotblat, 9 December 1964. RTBT 5/2/1/13 (43). At this meeting, COROIPAS became IPRA: both were peace organizations, with membership dominated by the social and economic sciences and promoted interdisciplinary research.

[34] PSGES Aide Memoire 1, December 1965. RTBT 5/2/4/1.

[35] PSGES Aide Memoire 1, December 1965. RTBT 5/2/4/1.

[36] Aide Memoire: written by co-chairs (Němec, Sojak, Vavo Hajdu, Liska (all Czechoslovakia), and Wilhjelm (Denmark), Valkenburgh (Netherlands), Klafkowski and Lapter (Poland), and Björnerstedt and Sparring (Sweden) Coordinated overall by Antonin Šnejdárek and Jens Adler (Czechoslovakian and Danish, respectively).

[37] A set of records relating to the PSGES is held in the Joseph Rotblat collection, Churchill Archive Center, University of Cambridge, UK. These records are substantial, but incomplete, precluding assembling a complete record of participation at each meeting. Participation lists for the PSGES are included as appendices in both books on the PCSWA by Joseph Rotblat but do not provide a breakdown of attendance at each meeting.

Table 5.3 Meetings of the Pugwash Study Group on European Security, December 1965–May 1968[39]

Meeting Number	Date	Venue	Funding
1 Inaugural	13–15 December 1965	Prague: Czechoslovakian Academy of Sciences	Czechoslovakian Academy of Sciences
2	1–5 March 1966	Hälsingborg, Sweden	The Carlsberg Foundation Via the Danish Pugwash Group
3	21–24 May 1966	Geneva Hotel Ambassador Switzerland	Private donation (Likely: James Wise)
4	7–8 September 1966	Jablonna, Poland	Polish Academy of Sciences
5	21–25 February 1967	Zagreb	Yugoslavian Academy of Sciences
6	13–16 May 1967	Marienbad Hotel Esplanade Czechoslovakia	Czechoslovakian Academy of Sciences
7	2–4 February 1968	University of Kiel, West Germany	Currently unresolved
8	13–18 May 1968	Marienbad Hotel Esplanade Czechoslovakia	Czechoslovakian Academy of Sciences

Soviet scientists attended only very occasionally—in striking contrast to the close involvement of both in the BW group (Kaplan 1999; Perry Robinson 1998).[38] The inertia of the superpowers towards the PSGES is interesting, especially perhaps in the American case, given Henry Kissinger's role in securing initial American funding for the project and his involvement in Working Groups 3 and 1 at Dubrovnik and Karlovy Vary respectively. Perhaps neither superpower had the appetite for involving themselves in Pugwash discussions on the German question. After all, this situation was created and sustained by their rivalry, and it remained arguably the most fraught and intractable political problem of the Cold War. Perhaps too they viewed the PSGES as one means to contain this issue and to keep it off the agenda at the international conferences.

Members of the PSGES wasted no time in getting to work. At the second meeting in Hälsingborg in March 1966, fifteen new reports had been written by members primarily on the German question, whilst several working groups—replicating the format of the Pugwash conferences—had been formed to tackle the different parts

---

[38] The incomplete records of the PSGES in the Rotblat Collection mean that the author cannot currently verify the extent of American participation in the PSGES. Julian Perry Robinson reports that Henry Kissinger was present at the sixth meeting of the group held in Marienbad in May 1967.

[39] Table based on sources in RTBT 5/2/4/1–8 which holds the records of the PSGES, but which are incomplete.

of its agenda.[40] The German-German relationship was an immediate priority for the PSGES, and in Hälsingborg it was agreed to "set up a committee of scientists of the two German states to study the mutual relations between them and a normalization of their relations."[41] Rotblat went to Hälsingborg where he met privately for discussions with Adler, Šnejdárek, and colleagues from both German states, Horst Afheldt, Harry Wünsche and Peck.[42] Indeed, there was enormous satisfaction at the active involvement of scientists from both Germanies, working, as Šorm put it in 1966, "side by side" in "an exceptional case in the sphere of international relations."[43] Scholars from both German states were highly active in the PSGES, and in February 1968 the seventh meeting of the group was held in Kiel—marking the first formal Pugwash meeting hosted by a German state.[44] In František Šorm's view, Hälsingborg proved the "viability" of the PSGES, but he was quick to emphasize that the Pugwash affiliation was its "greatest asset." If this was a strategic move to pay his dues to the Continuing Committee, it also reveals how the leaders of the PSGES saw the reputation of Pugwash on the international stage as a valuable resource.

Šorm would also have been aware of indications of early reservations about the PSGES project within the Continuing Committee. In January 1966, seven weeks after the inaugural PSGES meeting in Prague, Theodore Němec wrote to Rotblat, advising him that Ivan Málek had just updated the Czech Pugwash group that the Continuing Committee had formally approved the formation of the PSGES.[45] (Málek had been appointed to the Committee in 1964.) But, beyond this, Málek had also relayed to his Czech colleagues some disconcerting news about how, in contrast to the seemingly favourable stance towards the PSGES project at an earlier meeting of the Continuing Committee in December 1964, he had noted "a certain embarrassment or even coolness to this effort on the part of some of our members, particularly those from USSR and US."[46] Málek had confided to Němec that he found this odd, since the project had been discussed at that time with Khvostov and Henry Kissinger who had seemed "agreeable" towards it. Němec went on to ask Rotblat for his impression of the views of the Americans and Soviets on the PSGES, advising him that, in the meantime, Šorm would write to Millionshchikov, and Adler to Bernard Feld, to clarify the superpower stance on the new study group.

---

[40] PSGES Aide Memoire II, March 1966. PSGES 17. RTBT 5/2/4/2. The topics of the Working Groups included: Lawyers on a peace treaty with Germany; Problems of German reunification; Security; Integrational processes and cooperation between European states.

[41] František Šorm to Ivan Supek, 22 March 1966. (Between PSGES meetings 2 and 3) RTBT 5/2/4/3(2).

[42] PSGES Aide Memoire II, March 1966. PSGES 17, 3. RTBT 5/2/4/2.

[43] Šorm to Supek, 22 March 1966. RTBT 5/2/4/3(2).

[44] Documents relating to the PSGES meeting in Kiel, February 1968. RTBT 5/2/4/7. The Kiel-based West German lawyer Eberhard Menzel was a prominent figure within the PSGES and he had organized the PSGES meeting at the University of Kiel. Papers of Eberhard Menzel, Landesarchiv, Schleswig–Holstein.

[45] Minutes of the twenty-third meeting of the Continuing Committee, Addis Ababa, January 1966. RTBT 5/3/1/2 (5).

[46] Němec to Rotblat, 27 January 1966. RTBT 5/2/17/26.

This apparent shift in attitude amongst the American and Soviet members of the Committee towards the PSGES as detected by Málek and reported by Němec to Rotblat raises questions about whether this signaled concerns about the PCSWA engaging at a wholly new level with the politically incendiary German question. Perhaps this indicated an awareness that the PSGES could potentially make difficulties for the Pugwash organization as a whole, especially in terms of its relations with Washington—recently soured by Karlovy Vary, and with tensions over the Vietnam War simmering in the background. Perhaps internal matters were also in play, for example, the leadership may already have been wary of the zeal and boldness of the group, and/or worried about their ability to exercise control over it. Indeed, Němec's letter to Rotblat indicates that leading figures within the PSGES conceived its relationship to the Continuing Committee very differently than did some senior members of the Committee.

For his part, Rotblat was perhaps caught on the horns of a private dilemma. He had always held to the view that the primary focus of the PCSWA was and should remain nuclear disarmament: perhaps he was ambivalent about moving into overtly political terrain, especially where scientific expertise had little role to play. But, ever the diplomat and always pragmatic, he recognized the importance of the PSGES to the Europeans, who were now an influential and valuable constituency within Pugwash. Nevertheless, the shared reservations about the PSGES on the part of both the Americans and the Soviets within the Continuing Committee were telling, and may have cast a shadow over it from very early on.

The creation of the PSGES was a huge, pragmatic and perhaps inherently risk-laden step for the PCSWA. As an early means of tracking its development, the co-chairs were asked by Rotblat on behalf of the Continuing Committee to produce a report on their work for the 16th conference held in Sopot, Poland, in September 1966.[47] The theme in Sopot, 'Disarmament and World Security, especially in Europe,' resonated with the remit of the PSGES. Whilst this theme indicated a willingness on the part of the Committee to support stronger engagement with European security—it decided on conference topics—'European security' was framed in the wider context of 'World Security'. This point was driven home by Bernard Feld and Rotblat in Sopot (Feld 1966). In his opening address Rotblat, whilst acknowledging the need to engage with European security, restated that the priority for PCSWA was and remained the nuclear arms race and nuclear proliferation—noting that on the day the conference began, the French were conducting another atmospheric nuclear weapons test at its test site at Ekker in Algeria (Rotblat 1966).

5.3 Early Signs of Trouble: The Issue of PSGES Autonomy

If Sopot represented a major flashpoint in relation to the issue of the Vietnam War, it was also significant for the PSGES. The group's report, co-written by the group's chairmen, Jens Adler and Antonin Šnejdárek, worried the Secretary General and

---

[47] Likewise, the same was asked of the Biological Weapons group.

some of his colleagues on the Continuing Committee.[48] If the pace, intensity and scope of the work of the PSGES perhaps came as a surprise to them, the ambitious plans set out in the report likely sent ripples of unease within the beleaguered Committee. Up and running for eighteen months, the group had already met four times: the fourth meeting took place in Jablonna, at the invitation of the Polish Academy of Sciences, just prior to the Sopot conference. The PSGES report detailed progress in its program of work organised around its three priorities: Security, Germany, and Integrational Processes (political, economic and scientific cooperation in the European region). A fourth section set out plans for Future Activities that included calling conferences, arranging seminars and, most immediately, another meeting to be held in Zagreb in February 1967—to which it would invite "as observers" four members of the Continuing Committee, including the Secretary General. Other plans included working together with SIPRI and IPRA on large research projects. The co-chairs reported too that they were exploring the possibility of having "governmental representatives from each of the four previous powers of allied occupation of Germany to give seminars on the points of view of their respective governments concerning European affairs, starting out with representatives from France."

The report revealed its members to have already produced a considerable corpus of research, citing over thirty research papers and it also noted that several Working Groups were up and running. These research papers were circulated amongst the members of the PSGES and sometimes shared also with colleagues beyond the group, for example, with institutes of international and economic affairs. This seemingly casual practice of sharing Pugwash documents outside the organization was of particular concern within the Continuing Committee. Furthermore, there were references to "not binding" the Continuing Committee to these reports, indicating an intent to act ever more autonomously. In their covering letter, Adler and Šnejdárek also proposed that in moving beyond "the initial and necessarily experimental period of work," the size of the meetings might be increased: they indicated too that in future they would like to work with the Committee on compiling the invitation list.[49] At least some of these practices, the implicit assertion of a degree of autonomy, and the bold agenda would have struck a wrong note—if not set alarm bells ringing—within the Continuing Committee.

Overall, the report set out a wide-ranging program of work that was extending in multiple directions, including beyond Pugwash, and strongly gave the impression of an independent project building links with policy, political, legal and government actors in different countries and with a range of NGOs. This ran contrary to the Pugwash tradition of remaining wholly independent. The PSGES was also asserting control over its publications—hitherto, the jealously guarded prerogative of the Continuing Committee. In short, the report made clear that the PSGES

---

[48] David Jens Adler and Antonin Šnejdárek. Report on the work of the PSGES, December 1965 to July 1966. PSGES-35. RTBT 5/2/17/25.

[49] Adler and Šnejdárek, Covering letter to the Continuing Committee, 27 July 1966. PSGES-35, Appendix 3. RTBT 5/2/17/25.

was taking decisions with far-reaching implications with little regard for Pugwash tradition. Moreover, it clearly indicated a pattern whereby PSGES decisions and actions preceded consultation with the Continuing Committee and/or the Secretary General—which were informed only in retrospect. This contrasted sharply with practice within the Biological Weapons group which, as was clear in its report to the Sopot conference, consistently worked through the Committee as it developed its program of work.[50] Leading figures within the PSGES were consulting mainly with each other to create, in effect, an independent forum of exchange and discussion, creating a novel network and identity as it formulated policy and strategy for the group. In by-passing the Committee, the PSGES leadership was subverting the traditional chain of command within Pugwash.[51] In addition to moving well beyond traditional Pugwash territory, the PSGES was adopting modes of working that contradicted rules—laid down, of course, informally—that had long defined the way of working within the Pugwash organization. The power relations underlying the informal hierarchy over which Rotblat and the Committee had hitherto presided and which they saw as essential to the international role and work of Pugwash were seemingly being challenged from within.

Rotblat and some of his colleagues in senior Pugwash circles may also have harbored another nagging doubt about the PSGES, one relating to its expertise profile. The political problems integral to European security called for expertise other than that of physics and the other natural sciences. Hitherto, the Pugwash organization had built its identity around scientific and technical expertise—a unique characteristic that had been the basis of its relevance to governments. The PSGES departed from this tradition. A handful of physicists and chemists aside, the group was dominated by the social, human, political and legal sciences (Olšáková 2018, 2020; Isaac 2007; Solovey and Cravens 2012; Cohen-Cole 2014). Certainly, it would not have gone unnoticed by the Continuing Committee that science and technology barely featured in discussions within the PSGES: to its critics, the group had lost sight of the raison d'etre of the PCSWA.

For the moment, the PSGES continued apace with further meetings scheduled for February 1967 in Zagreb and, taking advantage of a standing invitation from the Czech Academy of Sciences, in Marienbad in May. But the group was by now a frequent topic of discussion at meetings of the Continuing Committee, including that held in Sopot in September 1966:

> In discussion, there was general dissatisfaction with the way the PSGES was developing, particularly the proliferation of its membership and lack of specificity in the topics discussed. It was felt that if (it) is to continue it should have a relatively small but technically effective membership and concentrate more on specific problems which may yield results. It was agreed that a change in the organization may be necessary.[52]

---

[50] Report of the Biological Weapons Study Group. *Proceedings of the 16th Pugwash Conference, Sopot, September 1966:* 88–105.

[51] For example: Frantisek Šorm to Ivan Supek, 22 March 1966; Günther Rienäcker to Anton Šnejdárek, 14 April 1966. RTBT 5/2/4/3 (2).

[52] Minutes of the twenty-fourth meeting of the Continuing Committee, Jablonna/Sopot, September 1966. RTBT 5/3/1/2 (5).

There appears to have been a further concern among western members of the Continuing Committee that the PSGES was becoming a powerbase of East European scholars from the legal and economic sciences. It remains unclear as to how the Soviet members of the Continuing Committee viewed the PSGES and/or influenced the work and role of their Czechoslovakian colleagues, who were taking an increasingly prominent role within it. What was clear was that in angry exchanges at Sopot over Vietnam, the Czechoslovakian Pugwash group, notably Theodor Němec, lent vocal support to their Soviet colleagues leading the attack on the US. Eastern bloc solidarity was always at work within Pugwash—as was its Western counterpart.

Whilst PSGES members enthusiastically went about their work, in early April 1967, the co-chairs of the group, Jens Adler and Antonín Šnejdárek, met with Rotblat in London, where plans for its upcoming sixth meeting in Marienbad were discussed.[53] As a result, some changes were made to the agenda for this meeting, for example, the Non-Proliferation Treaty (NPT) was added, as was East–West cooperation in "Big Science" projects. The co-chairs had clearly received a 'steer' from Rotblat to put nuclear disarmament on the agenda, and to include more discussion of science and technology related topics.

Three weeks later the Secretary General received a letter from Adler setting out his concerns about the group.[54] Adler reported that he had been pleased with decisions taken at the third PSGES meeting in Geneva in May 1966 where a "Structure and method" of work had been agreed, including that future meetings would focus on "more limited and concrete topics," that the number of participants be reduced drastically, and that participants be invited "only upon expertise criteria."[55] But Adler now reported that, to his "great disappointment," none of this had been in evidence in either Jablonna or in Zagreb in February 1967. Adler was apologetic about what was happening. He went on to confide in the Secretary General that,

> The very fact that the group (PSGES) in a most unfortunate manner has developed into a kind of pseudo Pugwash conference, that the number of participants is steadily growing, that the agendas of the meetings are becoming increasingly vague and broad instead of defined and limited, and that discussions during the meetings conform more with the diplomatic and political relations between "the countries" invited than with free and unbound, idea-generating, scientific discussion.[56]

In describing the PSGES as developing into a kind of "pseudo Pugwash conference," Adler was criticizing its growing autonomy, not least making plans to arrange its own conferences and seminars, and forging links with external organizations. But he also disliked the way in which its meetings were moving away from scientific discussion and how, in his view, the PSGES was developing a style of diplomacy that was overly formal, even official, altogether different in practice to the unofficial and informal *modus operandi* he had initially envisaged for it. Such were Adler's concerns

---

[53] Notes for the 6th meeting. The co-chairmen, PSGE-41. RTBT 5/2/4/6 (1). Who called this meeting and why remains unknown—it did not coincide with a meeting of the Continuing Committee.

[54] Adler to Joseph Rotblat, 22 April 1967. RTBT 5/2/4/6 (4).

[55] On this meeting, see: PSGES Aide Memoire 3, May 1966. RTBT 5/2/4/3.

[56] Adler to Joseph Rotblat, 22 April 1967. RTBT 5/2/4/6 (4).

that he suggested to Rotblat that the upcoming PSGES meeting in Marienbad be "the last one," adding that in future some of its topics might be dealt with more effectively in smaller "informally structured" events. Adler was also of the view that some topics taken up within the PSGES were inappropriate for Pugwash. Having spearheaded the creation of the PSGES in 1964–1965, Adler was now a disillusioned critic of it. Around this time, moves were afoot within the PSGES to replace Adler as co-chair because the group was in an intense phase of development and some felt that Adler was distracted by political events of "1968" in Denmark. The Brussels-based lawyer Robert Leclerq was proposed as Adler's successor.[57] Tellingly, this move soon came about, replacing the chemist Adler with a lawyer well-versed in international relations, experienced in Brussels bureaucracy and routinely moving in official diplomatic circles. Leclerq's skills and attributes would be useful assets for the PSGES in its work on processes of European integration and in any dealings with the European Economic Community.

The next meeting of the Continuing Committee in May 1967 was busy, with lengthy discussions about the PSGES and about the Pugwash response to the Non-Proliferation Treaty.[58] The Committee expressed "general dissatisfaction" with the group and agreed that "a change in the organization of the group may be necessary." It decided that as a priority this would be raised in Ronneby. "In the meantime," the Committee decreed that "no further meetings of the group are to be held." However, at the PSGES, plans were already well underway for two meetings in 1968: in February in Kiel and in Marienbad in May. On learning this, the Continuing Committee conceded "Not much else could be done,"—an indication of the liberal, 'gentlemen's agreement' culture prevailing within the PCSWA—and charged the Secretary General with keeping in touch with the group.[59] Indeed, in its difficulties with the PSGES, the leadership was facing a new situation. Challenges of this kind that went against the established ways of working within Pugwash were unprecedented. The Committee struggled to devise a means of handling the group, especially given its preference for informal modes of working that emphasized collegiality.

Adler's insights had clearly deepened frustration with the PSGES among some within the Continuing Committee—especially perhaps the Secretary General. There was particular annoyance that whilst benefitting substantially from the Pugwash name, the group was by-passing the leadership and behaving in ways that went against Pugwash tradition. Rotblat found himself in an invidious position. Beyond his concern to remain neutral as Secretary General, he was reluctant to rein in the PSGES because of his European colleagues' clear commitment to it. Moreover, to those involved, the group was productive and proving successful. In addition to

---

[57] PSGES Aide Memoire 7, Kiel, February 1968. RTBT 5/2/4/7.

[58] Minutes of the twenty-fifth meeting of the Continuing Committee, 25–27 May 1967. RTBT 5/3/1/2 (5).

[59] Minutes of the twenty-seventh meeting of the Continuing Committee, 10–11 December 1967, St. Bartholomew's Hospital, London. RTBT 5/3/1/2 (6).

divisions within the leadership over Vietnam, the Secretary General had also now to contend with another potential source of division along a different axis of the Pugwash organization. Indeed, that both these sources of internal conflict emerged at the same time placed great demands on the Secretary General. Rotblat's efforts to remain impartial in this role, and his strict interpretation of the PCSWA principle of political neutrality, created an uneasy platform from which to navigate and tackle the conflicts and tensions, open and less open, that by 1967 were tearing the organization apart.

5.4 Informal Dialogue on Disarmament: The Loss of the Pugwash Monopoly

But the PSGES and Vietnam were not the only problems facing the Pugwash organization. It was also facing competition from new institutional actors moving into the disarmament and security field. The arrival in this territory of other actors, not least, the SADS group in 1964 and the SIPRI organization in 1966, was eroding its once dominant position as a forum for informal East–West dialogue on disarmament. Moreover, the fact that the SADS group included natural scientists posed a double challenge to the PCSWA, arguably diminishing the value of its distinctive core asset, namely that of scientific and technical expertise. If Pugwash had successfully exported its unique model into another organization, it had created what was proving to be a strong competitor. Indeed, at the Karlovy Vary conference in 1964, the SADS co-chair Paul Doty, in reporting on the group's recent inaugural meeting, commented exactly on this issue:

> It is natural to wonder how the work of the East–West Study Group will relate to the work in the same area that goes on at Pugwash meetings. However, this extra channel of thought should make itself felt in the improved efficiency of work done at the Pugwash conferences.[60]

Doty was doubtless seeking to gloss any potential rivalry, and he left aside the thorny question as to whether the SADS group could create a vacuum within Pugwash. The topics on the SADS agenda mirrored those of the PCSWA and many members of the new group—including, on the Soviet side, Mikhail Millionshchikov, Vasily Emelyanov, Nikolai Talensky, Vladimir Pavlichenko, and alongside Doty, the Americans Jerome Wiesner, Donald Brennan, Franklin Long, Louis B. Sohn and Henry Kissinger—were, in this period, regulars at the PCSWA.[61] There may have

---

[60] Report on second plenary session, 13 September 1964. East–West Study Group, chair: Bentley Glass. Report by Doty and Millionshchikov. *Proceedings of the 13th Pugwash Conference, Karlovy-Vary, September 1964*: 71–73.

[61] Others involved in SADS included: from the USSR, L.I. Sedov. From the US: D.H. Frisch, M. Shulman, George Kistiakowsky, Carl Kaysen, Mrs. B. Lall and, for a short time, F. Fletcher, Isador

been synergies, but the impression of overlap, duplication and potential rivalry was unavoidable. Doty and, until his death in 1973, Millionshchikov, remained influential in both the PCSWA and the SADS group—using and benefiting from both as they worked for disarmament.[62] Millionshchikov—regarded by Doty as 'more western' than his colleagues in outlook—emerged as the pivotal player on the Soviet side within the SADS group especially in its work towards the ABM treaty (Kubbig 1996, 25–40).

As Matthew Evangelista has noted, in the 1960s Cold War, the number of transnational actors rose markedly so that the international stage grew increasingly crowded (Evangelista 2010). This was especially true of the disarmament field, which was increasingly coupled to security. SIPRI provides a case in point. Formed in 1966 and with close relations to Pugwash—Rotblat was for many years on its Board of Governors—SIPRI rapidly established a formidable reputation within the 'disarmament and security' research field, including with regard to chemical and biological weapons—where it worked closely with the Pugwash Biological Weapons Study Group (Blackaby 1996).[63] Pugwash was in danger of being outflanked by the competition, especially the SADS group, SIPRI and the Dartmouth Conferences (Blackaby 1996; Vorhees 2002). The irony of the situation did not pass Rotblat by, as he observed in April 1967, it seemed as if "the very success of Pugwash in stimulating research and contacts, and in contributing to the relaxation of tensions, is responsible for our decline."[64]

Nevertheless, time and again, the Continuing Committee—and especially the Secretary General—reiterated that the primary focus of the organization remained disarmament. To be sure, discussions about disarmament continued at the PCSWA throughout the 1960s, especially within Working Groups. Indeed, important contributions were made at the conferences, notably the paper discussing ABM by the American physicists Jack Ruina and Murray Gell-Mann at the twelfth PCSWA held in Udaipur, India, in early 1964. Ruina and Gell-Mann's paper proved controversial and ignited wider US-Soviet discussions on this topic, including within the SADS group (Kubbig 1996, 25–40; Ruina and Gell-Mann 1964). Meanwhile, Pugwash conferences provided regular opportunities for conversations between SADS members, as was the case in Dubrovnik in 1963 (Kubbig 1996). But the existence of the SADS group likely had implications for the standing in Washington of Pugwash as a forum for informal bilateral US-Soviet disarmament conversations. The loss of the Pugwash monopoly as an international venue for informal dialogue

---

Rabi and Jack Ruina. Report on East–West study group by Paul Doty and Mikhail Millionshchikov, PCSWA 13, *Proceedings of the 13th Pugwash Conference, Karlovy-Vary, September 1964:* 71–73.

[62] Paul Doty was also active in the Pugwash Biological Weapons Study Group.

[63] Solly Zuckerman served on the early SIPRI Governing Board and Scientific Council—as did Henry Kissinger, Carl Kaysen and Alistair Buchan. SIPRI benefitted from stable and generous funding, most prominently, from the Swedish government.

[64] Joseph Rotblat 1967. Memorandum: Future of Pugwash, 2. RTBT 5/3/1/19.

about disarmament fed into the crisis engulfing the PCSWA and added further to a sense that the organization was losing direction and momentum.

5.5   Whither Pugwash?

In spring 1967, Rotblat felt the situation to be so serious that the future of the PCSWA was in doubt, and he circulated a confidential memo about what he saw as a serious crisis to some of his senior colleagues.[65] Highlighting the very different institutional landscape and political environment in which Pugwash was now operating, he warned that the organization was losing influence with governments. It is also clear that he was weary with the divisions caused by Vietnam:

> We have lost our impact. … The enthusiasm has gone out of the national groups, the majority of which are hibernating. Governments seem to have lost interest in us, and where there is an interest it is often negative: in the socialist states we are criticized for not taking up stands on topical issues, such as condemnation of US aggression in Vietnam; in the western states we are criticized for following too much the policies advocated by the socialist camp.[66]

Rotblat was clearly worried that Pugwash was losing focus, momentum and status with all of its audiences. Painting a bleak picture of decline—apparent in his frustrations with what he saw to be the under-performance of some of the national groups—he was also concerned about the changing disciplinary character of the conferences, which now included "an increasing proportion of social scientists, which—coupled with the greater emphasis on political rather than technical aspects of the problems discussed—makes understanding and agreement more difficult."[67] This was surely a reference to the PSGES.

But Rotblat also identified a number of systemic problems. For example, he called for a stronger infrastructure, outlining an ambitious list of proposals including increasing by ten-fold the number of scientists active in the PCSWA, the creation of a new body separate from the Continuing Committee with responsibility for "carrying Pugwash findings to the wider public and taking a public stand when necessary," and the enlargement of the office of the Secretary General to include a salaried staff. He lamented the over-reliance of the PCSWA on a handful of Pugwash veterans and called for abandoning the practice at conferences of having "official messages from dignitaries"—two features which he saw as implicated in a view of the PCSWA as an "exclusive club."[68] Rotblat linked this to the organization's troubled relationship with the wider scientific community, arguing that many colleagues viewed Pugwash as a "waste of time," and that this was contributing to its difficulties in attracting

---

[65] Joseph Rotblat. Memorandum: Future of Pugwash. 1967. RTBT 5/3/1/19. Rotblat first registered concerns along these lines at the twenty-fourth meeting of the Continuing Committee held in September 1966 during the Sopot Conference. RTBT 5/2/1/3 (5).

[66] Joseph Rotblat. Memorandum: Future of Pugwash, 1967, 2. RTBT 5/3/1/19.

[67] Ibid 3.

[68] Ibid 5.

'new blood'—something he had long been concerned about.[69] Indeed, one place where new—and enthusiastic—Pugwash participants were to be found was within the PSGES—which can only have added to Rotblat's frustrations.

In part, this was a crisis of the generation of physicists who had played a role in the birth of the atomic age. At this time, the average age of those attending Pugwash conferences was fifty-nine. Of course, seniority typically came with age, but retaining seniority within its ranks across generational change meant attracting younger scientists into the Pugwash organization. But the upcoming generation of scientists attaining seniority did not relate to the nuclear threat in the same way as their predecessors. This generation had not participated in the Manhattan Project, and perhaps too some had learned to live with the bomb. Furthermore, in the 1960s, the work of Thomas Kuhn, C.P. Snow's *Two Cultures* and other sociological analyses of science, together with the rise of the new academic field of Science Studies, were critiquing the idea of science as neutral and its practitioners as impartial—challenging the ideas and values on which the PCSWA had been built (Agar 2006; Kuhn 1962; Snow 1962).

The crisis facing the Pugwash leadership was rooted partly in a loss of consensus. The consensus on which a decade earlier the PCSWA had been built (against nuclear weapons, and scientists' guilt about having developed these weapons) was dissolving amid the changing geopolitical and intellectual environment of the mid-1960s Cold War. This raised the critical and increasingly urgent question as to what would replace that consensus in the future. Establishing a new consensus posed difficulties since younger scientists viewed the problems facing the world—and their place and responsibilities in it—very differently to their predecessors. Moreover, the same was true for scientists beyond the US, Soviet Union and Europe. In short, the situation was very serious—leading Rotblat to propose that:

> It seems to me that if we are not prepared to take the drastic steps that are necessary to remedy the situation the honourable thing would be to go to the Ronneby Conference and declare frankly that in the changed situation in the world, and since we are not prepared to put in a greater effort, we cannot keep Pugwash going. We should then propose that some of our activities be handed over to other organizations, such as SIPRI, IPRA or ICSU, and recommend that the Movement be wound up.[70]

Perhaps Rotblat hoped that making this radical suggestion might galvanize his colleagues into action. However, it also makes clear the depth of his frustration. It was perhaps an indication that he was fatigued by the increasing demands that the PCSWA and his role as Secretary General were making on his time and energy. Indeed, elsewhere in his memorandum, he announced his intention to step down at the upcoming seventeenth and quinquennial conference in Ronneby, Sweden, in September 1967. In the event, he did not resign: he was persuaded to stay on as Secretary General, continuing in this role until 1973.

---

[69] For example, Rotblat raised this issue in his opening address in Venice. Statement of the Secretary General. *Proceedings of the 15th Pugwash Conference, Venice, April 1965:* 71–75.

[70] Joseph Rotblat. Memorandum: Future of Pugwash, 1967, 4. RTBT 5/3/1/19.

Rotblat was highlighting the need for change in order that the PCSWA keep its place, and its relevance, on the international stage. But he was also directly referring to the changed political context, most obviously superpower détente in the nuclear realm and, from this, the changing contours of the disarmament landscape in the wake of the LTBT, the imminent NPT, and intensifying discussions about the problem of anti-ballistic missiles, where the SADS group was becoming important (Kubbig 1996). Whilst welcomed by the PCSWA, détente and the NPT raised fundamental questions about its future role in the international disarmament conversation.[71] With nuclear disarmament treaties in place and others under negotiation, and with the existence of numerous respected channels for East–West dialogue on disarmament, what should the focus and priorities of the PCSWA be in the future? The PCSWA was not just being outflanked—it was also beginning to look out of step with the changing world of the mid-1960s Cold War.

Rotblat's senior colleagues also recognized the need for change. All wanted Pugwash to continue, but there was little agreement on future priorities and directions. Acknowledging that the "changed state of world affairs in the world" required a "review of the Pugwash program," Eugene Rabinowitch argued that in the future this had to include "a serious approach to the problems of development."[72] Meanwhile, the changing disciplinary background of Pugwash elicited much discussion and diverse responses. Hermann Bondi was against further diluting the Pugwash tradition as a stronghold of natural scientists, citing the oft-repeated narrative about the exceptional capacity of this professional group to work—and, significantly, build trust across—across national borders. As he put it: "I am not convinced of the wisdom of bringing in too many people from outside the natural sciences. I doubt whether they have that degree of professional cooperation across international boundaries that we have, and so they may well bring in an element less likely to induce the trust so essential."[73] In contrast, Vikram Sarabhai, by now a veteran of the Continuing Committee, argued against Rotblat's view that the inclusion of social scientists introduced weakness into the PCSWA. Indeed, Sarabhai argued the opposite, that the

> effectiveness of the Pugwash Movement in terms of real problems can only arise if we appropriately consider the multi-dimensional aspects of each issue. Political and social factors are undoubtedly important dimensions in disarmament or development. If our discussions have not been effective recently, I would suggest that this may be due to not having effectively involved social sciences of the right background and papers produced by them with appropriate exactitude.[74]

Rudolf Peierls concurred: "I think it is a good thing to bring in social and political scientists when so many of our problems are non-technical, and the knowledge and

---

[71] Statement on the NPT issued by the Pugwash Continuing Committee. Marienbad, 13–15 May 1967. RTBT 5/3/1/20–2. The statement began by urging the 18 Nations Disarmament Committee of the United Nations "to formulate and agree upon such a treaty with all possible haste."

[72] Eugene Rabinowitch to Joseph Rotblat, 24 April 1967, 2. RTBT 5/3/1/19.

[73] Eugene Rabinowitch, Thoughts on Pugwash. 28 April 1967. RTBT 5/3/1/19.

[74] Sarabhai to Rotblat. RTBT 5/3/1/19.

experience of these people may be very relevant." But, agreeing to an extent with Bondi, he also thought there were limits to this.[75]

All recognized that standing still was not an option. Wherever the leadership looked, it faced daunting challenges. The PCSWA had to respond to the changing geopolitical context in which it was operating. In 1967, the organization was approaching its tenth anniversary. The upcoming conference, scheduled for September in Ronneby, Sweden—and funded by SIPRI—was its second quinquennial conference and, by convention, an occasion for reviewing its work over the last five years and deciding on its future agenda and priorities.

5.6 Ronneby, September 1967: Reasserting Pugwash Tradition, Resetting the Agenda

The Ronneby conference took place between 3–8 September 1967 and involved 180 participants. Organized around the broad theme of "Science and World Affairs," the program involved seven plenary sessions and seven working groups.[76] Each morning was given to plenary sessions, including on the themes of: responsibilities of scientists; arms control, peacekeeping and security; new approaches in disarmament, and international cooperation and development.[77] Afternoons were given to the Working Groups, the themes of which included: 'Education, technology and development—use of nuclear power to increase resources, assessing technical assistance programs (bi/multi-lateral).' The final day was entirely a plenary session given to summaries and discussion of the work/findings of the Working Groups, the crafting of the conference statement, and the election of new members to the Continuing Committee. The Committee had taken steps in advance to contain in Ronneby the still-simmering tensions over Vietnam. The by now familiar strategy centered on compartmentalization, namely assigning discussion of the conflict to the seventh Working Group given to 'Current conflicts and their resolution,' with which, "it was agreed," that there will be no discussion of the Vietnam issue at plenary sessions."[78] As a quinquennial conference, Ronneby included two so-called Standing Committees—addressing respectively 'future activities' and 'future organization' which took a primary role in the review process. Membership of these committees included Pugwash veterans mixed with a sprinkling of senior colleagues who had perhaps attended one or two conferences and who were sympathetic to the organization. That is to say, the Standing Committees were of Pugwash rather than independent of it.

---

[75] Peierls to Rotblat, 25 April 1967, 1. RTBT 5/3/1/19.

[76] Proceedings of the 17th Pugwash Conference, Ronneby, September 1967: iv.

[77] Conference program, Ronneby, 1967. RTBT 5/2/1/17–49; PCSWA 17. Proceedings of the 17th Pugwash Conference, Ronneby, September 1967: iv–xi.

[78] Minutes of the twenty-fifth meeting of the Continuing Committee, 25–27 May 1967, 7. RTBT 5/3/1/2 (5).

The reports from both Committees were also discussed in plenary session on the final day.[79]

In the plenary sessions, Eugene Rabinowitch placed the current crisis facing the PCSWA center stage in his paper.[80] Acknowledging that many within the organization "cannot help feeling that it is somehow failing," that the whole enterprise might be "fizzling out," he reflected that this had become something of "marking time" period. He attributed the current difficulties in part to the wider political situation, as he put it: "The main reason for the Pugwash doldrums is the direction international relations have taken in recent years." He emphasized the need to ride out the vagaries of the current international situation, emphasizing too that—in his view— the PCSWA was a long-term project that could not control but had to learn to adjust to the changing dynamics of the Cold War.

In his address as Secretary General, Joseph Rotblat targeted the PSGES, remarking on its less than satisfactory progress in tackling the "very diffuse, mostly political, and highly controversial" problems of European security and comparing it unfavourably with the achievements of the BW Study Group.[81] If Rotblat was emerging as the key protagonist in tackling the PSGES problem, not everyone shared his views. Most notably, the report of Working Group 2, which addressed the theme 'Peacekeeping and Security,' gave a resounding endorsement of the PSGES, recommending that its program be implemented, even expanded, and moreover, that it "should be authorized to initiate specific research projects."[82] Unsurprisingly, the twenty-five man group— drawn from across the blocs—was stacked with PSGES stalwarts, including its co-chair Antonin Šnejdárek, East Germans Hans Kröger and Harry Wünsche, West Germans Eberhard Menzel and Ludwig Raiser and, interestingly, Jens Adler. Their report was strikingly at odds with Rotblat's views of the PSGES. These divergent views point to tensions—even perhaps an emerging power struggle—at the highest levels within Pugwash.

In hindsight, despite strong support for the PSGES in some quarters, its fate was likely decided at Ronneby. Important here was the Standing Committee on Future Activities. With a remit that included "to re-define briefly the raison d'etre of the PCSWA," including the scope of its activities and its priorities for the next five years, this Committee was chaired by the Scottish physicist Gordon Sutherland and

---

[79] Report of the Standing Committee on Future Activities, Ronneby, 1967. (Copy also in: RTBT 5/2/1/17 (51)). The question as to whether the Committee received a copy of Rotblat's 'Future' memo—or whether the Committee members learned of it by 'word of mouth'—is unclear. Report of the Standing Committee on Future Organization. The two reports are in: PCSWA 17. *Proceedings of the 17th Pugwash Conference, Ronneby, September 1967*: 195–199, 200–204.

[80] Eugene Rabinowitch, The rationale of Pugwash. The seventeenth PCSWA, Ronneby, Sweden, September 1967. RTBT 5/2/1/17.

[81] Joseph Rotblat. Report of the work of the Continuing Committee since 1962. 1967. RTBT 5/3/1/20 (2).

[82] Final Report of Working Group 2. Peacekeeping and security. PCSWA 17, Ronneby, Sweden, 1967. RTBT 5/3/1/20 (4). The other members of this group were: Louis B. Sohn (co-chair), Vladimir Aboltin, Adamczewski, Dobrosielski, Elm, Frank, Hajdu, Dorothy Hodgkin, James, Landheer, Moch, Muller, Munger, Parsons, Pochitalin, Sparring, Vavpetek, Voslensky and Yamada.

involved twelve Pugwash veterans—with the obligatory balance of Americans and Soviets, including for example Martin Kaplan and Mikhail Dubinin, and the Czech Theodor Němec.[83] The influential Kaplan was a leading figure in the BW group and a close ally of Rotblat; Němec was closely involved in the PSGES: it is hard to imagine that he did not speak up for it, but assuming that he did, he was overruled. In all likelihood, there were differences of opinion within this Committee about the future direction of the PCSWA, especially regarding the PSGES. For his part, Joseph Rotblat remained adamant that nuclear disarmament remained the absolute priority.

The report of this Standing Committee did not mention European security in its recommendations regarding the future priorities of the PCSWA. Rather, the report proposed that the scope of Pugwash lay in three main, broad areas: the prevention and cessation of wars, with an emphasis on disarmament; working towards an "acceleration of the improvement in the state of the less developed countries," and the promotion of international cooperation, including international projects in science, especially with regard to the countries of the Global South. In addition, a fourth general recommendation included that the PCSWA "promote a greater sense of social responsibility among all scientists regarding the political and social consequences of their scientific work." Tellingly, the contours of the report by the 'Future Activities' Standing Committee echoed the concerns set out by Rotblat in his 'Future' memo in the run up to Ronneby.[84] It is unclear whether members of this Standing Committee received a copy of the memo, or if they learned of it by word of mouth. At any rate, its report essentially called for a return to the scientific/technical focus of the PCSWA. This was seen as a means for the organization to retain its distinctive position within the expanding and diversifying international institutional landscape taking shape around a new research field encompassing, variously, disarmament, security, defense studies and peace-related research. As the Standing Committee put it,

> Although new organizations with closely related objectives have come into existence since 1957 the contributions which Pugwash can make to the solution of problems of world security and international co-operation are unique since scientists through their training and objective approach to problems should not only have a more detached outlook but a greater appreciation of the international possibilities (both dangers and opportunities) arising from scientific and technological developments.[85]

At one level, this was simply a restatement of the principle on which Pugwash had been founded, that scientific and technical expertise was its unique strength and the basis for its claims to be able to make contributions to disarmament. But on another level, this was setting out a "return to core" strategy for finding a way out of the crisis in which in 1967 the organization found itself.

---

[83] The Committee members were: Sir Gordon Sutherland (Chair; UK), Aklilu Lemma (Ethiopia), Rolf Björnerstedt (SWE), Harrison Brown (US), Mikhail M. Dubinin (USSR), R.V. Garcia (ARG), Petr L. Kapitza (USSR), Martin M. Kaplan (US), Michel Magat (FRA), Leo Mates (YUG), Theodor Němec (CZE), S.H. Zaheer (India).

[84] Standing Committee, Future Activities, Ronneby. RTBT 5/2/1/17 (51).

[85] Standing Committee, Future Activities, Ronneby. RTBT 5/2/1/17 (51).

Echoing further Rotblat's 'Future' memo, the Standing Committee report emphasized that the PCSWA had to bring new blood into the PCSWA and encourage greater activities on the part of the existing National Groups, whilst calling for the formation of new national groups in the "developing world." In tone, it also echoed the views of some within the Continuing Committee of the need for the PCSWA organization to move "towards professionalization and sharp focus."[86] Perhaps with this in mind, steps were taken to create a new role, that of Pugwash President, with a one year tenure. The first three holders of this post were John Cockcroft, Mikhail Millionshchikov and Eugene Rabinowitch—British, Russian and American respectively.[87]

The way out of the crisis that engulfed Pugwash in 1967 came to center on a return to a focus on science and technology, and reaffirmed the controlling influence of the Continuing Committee over all the activities of the international PCSWA. The report of the Standing Committee on Future Activities reflected/served the interests of the Continuing Committee in other important ways. It recommended that in future any/all statements in connection to the conferences "remain entirely in the control" of the Continuing Committee. Furthermore, meetings of both Study Groups and the "procedures for their operation," including organization and financing, were likewise to remain exclusively in its hands—a stipulation likely more palatable to the Biological Weapons (BW) group than to the PSGES. Likewise, the Continuing Committee was to remain as the sole adjudicator on publications arising from within the PCSWA, including the Study Groups, and over publicity about all Pugwash activities.

The Standing Committee on Future Activities also recommended a new kind of event: the Pugwash Symposium, each of which would focus on a specific topic. This was seen as a way to make more effective use of finite PCSWA resources, whilst adding simultaneously breadth and focus to its agenda. This new Symposia program would include several symposia each year, organized by national groups with each symposium being jointly convened by the (hosting) national group together with the Continuing Committee. It was hoped the Symposia could serve as one means to reinvigorate and boost participation in the national groups. The Continuing Committee had sole control over any publications arising from the Symposia.

The Symposia idea was taken up immediately: the inaugural Pugwash symposium took place in London in April 1968, convened by the British Pugwash group and addressing the theme of 'Control of peaceful uses of atomic energy.'[88] The Symposia came with a stringent set of rules formulated by the Continuing Committee which had exclusive control over issuing invitations and over any publications arising from these

---

[86] Continuing Committee meeting no. 25, May 1967, Geneva. RTBT 5/3/1/2 (5).

[87] Minutes of the twenty-ninth meeting of the Continuing Committee, Nice, France, September 1968, 15. RTBT 5/3/1/2 (6). Minutes of the thirty-first meeting of the Continuing Committee, Sochi, Soviet Union, October 1969. RTBT 5/3/1/2 (7). John Cockcroft's tenure was curtailed by his sudden death in 1967.

[88] For the full list of Pugwash Symposia, see RTBT 5/2/2/1.

international events.[89] The Symposia program, which went on to become a long-term mainstay of the PCSWA, reasserted the central place of science and technology on the Pugwash agenda. Significantly, in the early 1970s, this came to provide a convenient means to facilitate engagement with the countries and regions of the 'developing world.' Most immediately, for the Pugwash leadership, the Symposia came also to provide part of the solution to the PSGES problem.

5.7   1968: The End of the Road for the PSGES

The growing perception within senior leadership circles that the PSGES had somehow to be reined in translated into further action at the seventh PSGES meeting in February 1968 in Kiel, West Germany, organised by Eberhard Menzel, professor in international law at the city's university and prominent member of the PSGES.[90] Making an unusual appearance in any Study Group, Pugwash veteran Rudolf Peierls, recently appointed as chair of the Continuing Committee, relayed the news that the second Pugwash symposium was to take place in Marienbad in May 1968.[91] Jointly organized by the Czechoslovak Pugwash Committee and the Continuing Committee, this was given to the theme of 'Scientific and technical cooperation in Europe as a contribution to European security.' This was seemingly a reminder to the PSGES about the primacy of science and technology in the Pugwash project. This symposium, held between 13 and 18 May would coincide with the eighth meeting of the PSGES, also taking place in Marienbad.[92] It had been determined in Kiel that the

---

[89] Joseph Rotblat. Memorandum to Chairmen or Secretaries of National Pugwash Groups. 14 August 1968. RTBT 5/2/1/22 (2).

[90] PSGES, Aide Memoire VII. RTBT 5/2/4/7. Unusually, Carl Friedrich von Weizsäcker was present in Kiel, making a rare appearance at a Pugwash event—a reflection perhaps of the inclusion of the Non-Proliferation Treaty on the PSGES agenda which was highly controversial in the Federal Republic. He seems also to have attended the sixth PSGES meeting held in Marienbad, 13–16 May 1967. Carl Friedrich von Weizsäcker. Bericht über Pugwash Tagung Marienbad, May 1967. Records of the VDW, Bundesarchiv Koblenz, Bestand 456/342. Von Weizsäcker had long distanced himself from the PCSWA—much to the chagrin of the PCSWA leadership; see (Sachse 2016, 2018). The extent of West German concerns about the PSGES as a site of German-German dialogue/cooperation, and its work on the "German question" and the NPT remains to be clarified. Von Weizsäcker was at this time pursuing his own project of establishing a new forum for the study of the problems raised by science and technology under the aegis of the Max Planck Society.

[91] PSGES, Aide Memoire VII. RTBT 5/2/4/7.

[92] Timetable for second Pugwash symposium. RTBT 5/2/4/8 (2). Confusion surrounds this symposium: the minutes of the twenty-eighth meeting of the Continuing Committee, held in Marienbad between 16–18 May 1968, report that due to "organizational difficulties" and "dissatisfaction" with how things had gone in Marienbad, it was agreed with the French Pugwash group that a second part of the symposium would take place at the next Pugwash conference, to be held in September (1968) in Nice, France. RTBT 5/3/1/2 (6), p. 2. For records of the Marienbad (Part 1) and Nice (Part 2) symposia see, respectively, RTBT 5/2/4/8 and RTBT 5/2/2/3.

Fig. 5.1 Session of the Pugwash Study Group on European Security. Eighth meeting, Marienbad, 13–18 May 1968. *Source* RTBT 5/2/4/8 (5)

two events were to take place "quite separately but simultaneously," with Peierls making it clear that the symposium was "taken by the Continuing Committee as one action belonging (to the symposia series) and that it was not considered to be directly connected with the meeting of the PSGES." According to Peierls, this event would center on a paper to be prepared by Frantisek Šorm echoed the title as the symposium, namely, 'The scientific cooperation among European nations as an important precondition of the European security.'[93] The available sources do not clarify how these rather odd arrangements were arrived at, or give a sense of the mood in which they were agreed upon. At any rate, the message embedded within this move would not have been missed by Šorm—or anyone else.

In hindsight, the scheduling of the second symposium to coincide with the eighth PSGES meeting marked the end of the road for the PSGES: this meeting in Marienbad was its last. (Figs. 5.1 and 5.2). The similar themes of both meetings indicated that a transition was underway whereby discussions about European security would henceforth take place within the Symposia program. In practice, this provided a mechanism by which the PSGES was phased out and Pugwash engagement with European security limited to the Symposia. Whilst topics related to European security featured prominently on the early Symposia program, crucially they were always framed in ways that foregrounded scientific and technical aspects.[94]

The available evidence points to the demise of the PSGES being the outcome of a power struggle within Pugwash between, on the one hand, Europeans—especially Eastern Europeans, and on the other hand, the Secretary General and some members

---

[93] PSGES, Aide Memoire VII, 6. RTBT 5/2/4/7.

[94] Records of the Pugwash Symposia. RTBT 5/2/2.

Fig. 5.2 Joseph Rotblat and Academician Ivan Málek in conversation in the Town Hall Wine Cellar in Jachymov during the eighth meeting of the Pugwash Study Group on European Security, Marienbad, 13–18 May 1968. *Source* RTBT 5/2/4/8 (5)

of the Continuing Committee. The minutes of its meetings around this time relay, in typical cursory fashion, that the Symposia constituted "a better way of carrying on the debate on European problems" and that this had been communicated to the co-chairmen of the PSGES.[95] These sources reveal little regarding discussions about, let alone dissent and conflict over, the final meetings of the PSGES or the transition from PSGES to the Symposia.

At this time, members of the Czechoslovakian Pugwash group—including those active in the PSGES—were caught up in the unfolding national crisis. Indeed, any opposition to the demise of the PSGES was perhaps forestalled by geopolitical developments, specifically by the shocking events in Czechoslovakia during 1968 which all but destroyed the Czech Pugwash group thereby depriving the PSGES of a great deal of its driving force.[96] These events precipitated a shift that saw the Polish Pugwash group replace their Czechoslovakian colleagues as the favored satellite of the Soviet Union within the PCSWA (Olšáková 2018, 2020). At any rate, these developments may have both eclipsed and eased the process through which the Symposia replaced the PSGES as the primary forum within Pugwash for dealing with European security. (Fig. 5.3).

Characteristically, Joseph Rotblat would later 'gloss' this transition by framing the Symposia as a continuation of the PSGES—although he steadfastly remained

---

[95] Minutes of the twenty-eighth meeting of the Continuing Committee, 16–18 May 1968, Marienbad. RTBT 5/3/1/2 (6). Minutes of the twenty-seventh meeting of the Continuing Committee, 10–11 December 1967, St. Bartholomew's Hospital, London. RTBT 5/3/1/2 (6).

[96] Rotblat would later assert that for some Czech scientists their work with Pugwash had cost them dearly career-wise. Rotblat, Joseph. The Pugwash Movement in Europe. 1975. RTBT G75.

Fig. 5.3 Academician Ivan Málek, Cecil F. Powell and Eugene Rabinowitch in conversation during the fifth Pugwash Symposium, Marienbad, 19–24 May 1969. *Source* RTBT 5/2/2/5 (17)

quietly critical of the group.[97] For example, in his 1972 chronicle of the PCSWA, he remarked that:

> By the end of the third year the range of topics that had arisen in the discussions of the Study Group had grown so wide that it was difficult to maintain cohesion, and the Continuing Committee decided that the objects of the Study Group might be better achieved by organizing several Symposia, each concerned with one aspect of European security (Rotblat 1972, 29).

And in a 1975 lecture, he nodded towards how the differing expertise at the PSGES was implicated in the difficulties posed by the group and its 'fit' within the Pugwash organization:

> A characteristic feature of the Study Group was that the majority of participants were from the social sciences, mainly law and economics, unlike the other Pugwash activities in which natural sciences predominated. This was probably due to the fact that the topics discussed were much more of a political nature (…) This is not intended as a censure of the Study Group, which did some very useful work in laying the ground rules for the European Security Conference, but the fact is that the rapid increase in the number of participants and the widening range of topics discussed made the work of the Study Group less effective.[98]

Of course, leaders of the PSGES might have held a different view of events. Indeed, as Rotblat acknowledged, from "time to time" there were calls, "largely from social

---

[97] Ibid.
[98] Ibid.

scientists," for the group to be reconvened—but this did not happen.[99] Engaging with the political problems of European security on a study group basis—which, moreover, was avowedly interdisciplinary—proved a step too far for the Pugwash leadership. The problems and eventual abandonment of the group underlined the centrality of natural science expertise to the Pugwash organization: this was and remained the basis of its identity and reputation, and the framework for its engagement with political issues. The PSGES experiment determined the limits beyond which the PCSWA leadership would not go in terms of engaging with the purely political problems of the crisis engulfing Europe in the 1960s Cold War.

The demise of the PSGES reveals new insights into the internal power relations at work within the Pugwash organization. The founding and continuing preference of the leadership for informal modes of working was—so the narrative went—because this conferred on Pugwash a versatility important for its ability to mount a rapid response to external political crises, such as Cuba or Vietnam. However, this *modus operandi* served also to keep power firmly concentrated in the hands of the Continuing Committee. Paradoxically, for all the dissenting credentials of the western leadership of the PCSWA, there were limits to which deviating from the codes, rules and guidance set down informally by that leadership would be tolerated. The end of the road for the PSGES signaled that the leadership would wield its power and act not only to protect Pugwash tradition but to preserve and promote its vision of the organization. Above all, the story of the PSGES experiment also reaffirmed that science and technology remained the prism through which Pugwash viewed the changing Cold War world and remained the basis for its engagement with world affairs.

References

Agar, Jon. 2006. What happened in the sixties? *British Journal for the History of Science* 41 (4): 567–600.

Albrecht, Stephan, Hans-Joachim. Bieber, Reiner Braun, Peter Croll, Henner Ehringhaus, and Maria Finckh. 2009. *Wissenschaft—Verantwortung—Frieden: 50 Jahre VDW*. Berlin: Berliner Wissenschaftsverlag.

Blackaby, Frank. 1996. *SIPRI. Continuity and change 1966–1996*. 30th anniversary commemorative volume. Solna: SIPRI.

Bull, Alan T. 2011. Ivan Málek: A tribute. *Journal of Technical and Biotechnology* 86 (5): 621–624.

Cohen-Cole, Jamie. 2014. *The open mind: Cold War politics and the sciences of human nature*. Chicago: University of Chicago Press.

Evangelista, Matthew. 2010. Transnational organizations in the Cold War. In *The Cambridge History of the Cold War*, eds. Melvyn Leffler, and Odd Arne Westad, 400–442. Cambridge: Cambridge University Press.

Feld, Bernard T. 1966. General remarks on current problems of disarmament and European security. In *Proceedings of the 16th Pugwash Conference, Sopot, September 1966*: 71–75.

---

[99] Meanwhile, in 1969, the Biological Study group went into what Julian Perry Robinson had called a period of "abeyance"—but for very different reasons and in an amicable process agreed between the Continuing Committee and the leaders of this study group. This group was resurrected in the 1980s.

Green, John. 2017. *A political family. The Kuczynskis, fascism, espionage and the Cold War*. London: Routledge.

Isaac, Joel. 2007. The human sciences in Cold War America. *The Historical Journal* 50 (3): 725–746.

Kaplan, Martin M. 1999. The efforts of WHO and Pugwash to eliminate chemical and biological weapons—a memoir. *Bulletin of the WHO* 77 (2): 149–155.

Kraft, Alison. 2020. Confronting the German problem: Pugwash in West and East Germany, 1957–1964. In *Science, (anti-)communism and diplomacy: The Pugwash Conferences on Science and World Affairs in the early Cold War*, eds. Alison Kraft, and Carola Sachse, 286–323. Leiden: Brill.

Kubbig, Bernd W. 1996. *Communicators in the Cold War: The Pugwash Conferences, the U.S.-Soviet study group and the ABM treaty. Natural scientists as political actors: historical successes and lessons for the future*. PRIF Reports No. 44. Frankfurt am Main: PRIF.

Kuhn, Thomas. 1962. *The structure of scientific revolutions*. Chicago: Chicago University Press.

Maddrell, Paul. 2005. The scientist who came in from the cold: Heinz Barwich's flight from the GDR. *Intelligence and National Security* 20 (4): 608–630.

Mastny, Vojtech. 2008. The 1963 nuclear test ban treaty. A missed opportunity for détente? *JCWS* 10 (1): 3–25.

Olšáková, Doubravka. 2018. Pugwash in Eastern Europe: The limits of international cooperation under Soviet control in the 1950s and 1960s. *JCWS* 20 (1): 210–240.

Olšáková, Doubravka. 2020. Czechoslovak ambitions and Soviet politics in Eastern Europe: Pugwash and the Soviet peace agenda in the 1950s and 1960s. In *Science, (anti-)communism and diplomacy: The Pugwash Conferences on Science and World Affairs in the early Cold War*, eds. Alison Kraft, and Carola Sachse, 259–285. Leiden: Brill.

Perry Robinson, Julian. 1998. *Contribution of the Pugwash movement to the international regime against chemical and biological weapons*. Background paper. Pugwash meeting no. 242, 10th Workshop of Pugwash Study Group on the Implementation of the Chemical and Biological Weapons Conventions.

Rotblat, Joseph. 1966. Address of the Secretary General, Sopot, 1966. In *Proceedings of the 16th Pugwash Conference, Sopot, Poland, September 1966*.

Rotblat, Joseph. 1967. *Pugwash—A history of the conferences on science and world affairs*. Prague: Czechoslovak Academy of Sciences.

Rotblat, Joseph. 1972. *Scientists and the quest for peace. A history of the Pugwash Conferences*. Cambridge: MIT Press.

Rubinson, Paul. 2020. American scientists in "Communist Conclaves." Pugwash and anti-communism in the US. In *Science, (anti-)communism and diplomacy: The Pugwash Conferences on Science and World Affairs in the early Cold War*, eds. Alison Kraft, and Carola Sachse, 156–189. Leiden: Brill.

Ruina, Jack and Murray Gell-Mann. 1964. Ballistic missile defence and the arms race. In *Proceedings of the 12th Pugwash Conference, Udaipur, January 1964*: 232–235.

Sachse, Carola. 2016. *Die Max-Planck-Gesellschaft und die Pugwash Conferences on Science and World Affairs (1955–1984)*. Berlin: Max-Planck-Institut für Wissenschaftsgeschichte.

Sachse, Carola. 2018. The Max Planck Society and Pugwash during the Cold War: An uneasy relationship. *Journal of Cold War Studies* 20 (1): 170–209.

Schrafstetter, Susanna. 2004. The long shadow of the past: History, memory and the debate over West Germany's nuclear status, 1954–1969. *History and Memory* 16 (1): 118–145.

Snow, Charles Percy. 1962. *The two cultures and the scientific revolution*. Cambridge, UK: Cambridge University Press.

Solovey, Mark, and Hamilton Cravens. 2012. *Cold War social science: Knowledge production, liberal democracy, and human nature*. New York: Palgrave Macmillan.

Trachtenberg, Marc. 2010. The structure of great power politics. In *The Cambridge History and of the Cold War*, ed. Melvyn Leffler, and Odd Arne Westad, 482–505, 1963–1975. Cambridge: Cambridge University Press.

Voorhees, Jan. 2002. *Dialogue sustained. The multilevel peace process and the Dartmouth Conference*. Washington D.C.: US Institute of Peace Press.

Chapter 6
Conclusion

Abstract This chapter sums up the three main arguments of the book. Each involves a paradox, testimony to the many contradictions at the heart of the PCSWA in this period. First, its innovative mode of techno-political communication enabled Pugwash to reconcile the contradictory aims of working with state actors while challenging them, and helped its scientists to pioneer informal exchanges of the kind that came to be known as Track II diplomacy. Second, the commitment to political neutrality underpinned Pugwash's credibility in East and West, facilitated unofficial diplomacy and was key to the role and authority of the Secretary General. But political neutrality divided the organization internally when the western leadership refused to take a public stand over the Vietnam War. The third main argument posits that for all the emphasis placed on its informal modus operandi, the Pugwash organization was in practice deeply hierarchical, a paradox exemplified in the history of the PSGES.

Keywords Techno-political communication · Track II diplomacy · Working Groups · Political neutrality · Vietnam War · The German question · Pugwash Secretary General

In focusing on the staging, the content and the consequences of conferences held between 1960 and 1967, this book casts new light on the development of the Pugwash organization, on the transition from its dissenting origins to a respected interlocutor in East–West diplomacy, its internal dynamics, its changing agenda and how it was shaped by the political twists and turns of the Cold War. It makes clear that each of these conferences was very different: congeniality in Moscow turned to acrimony in Stowe, whilst the Vietnam War transformed the conferences in Venice and Sopot into proxy battlefields of the superpower rivalry. For the PCSWA, this period saw some important successes, notably the breakthrough in East–West dialogue achieved in Moscow, which revitalized the organization and took its scientists into the realm of informal diplomacy. There were also disappointing setbacks, notably the divisions caused by Vietnam, and the difficulties with the PSGES. The analysis of the

A. Kraft, *From Dissent to Diplomacy: The Pugwash Project During the 1960s Cold War*, SpringerBriefs in History of Science and Technology, https://doi.org/10.1007/978-3-031-12135-7_6

conferences makes clear the paradox of Pugwash as both a Cold War bridge and battleground between East and West. It shows also how profoundly the organization was affected by external geopolitical events.

Each chapter points to the fundamental importance of personal relationships within Pugwash to its work and its resilience. Through the sequence of conferences featuring in the different chapters, regular attendees became bound together by the shared experience of contributing to an evolving project. Nowhere was this more apparent than within the Continuing Committee which, as this book argues, was key to the durability of the Pugwash project. The Committee was engaged in a precarious balancing act, constantly looking both East and West as it guided the development of the conferences and steered the organization into and through new territory while striving also to meet the challenges arising from external political developments. Here, the importance of the bonds between the founding cohort of Pugwash scientists, within and across the blocs, cannot be overstated. These bonds stemmed from a strong sense of generational identity, rooted in the shared wartime experiences of this particular cohort of physicists coupled to their sense of unease about their role in bringing about the atomic bomb. As Joseph Rotblat put it, "We scientists have a great deal to answer for" (Rotblat 1998). For some within the founding cohort, a generational affinity was strengthened by their shared Jewish heritage and/or émigré backgrounds. These shared identities, cultural ties and common intellectual experiences created an ethos that served within Pugwash as a source of cohesion and resilience. But it was above all the scientists' shared concerns about the immediate dangers of the nuclear stand-off between the superpowers, and a conviction that they must act to avert them, that formed the consensus on which the PCSWA was built.

The particular affinity that the Pugwash founders felt between themselves as physicists was something they sought to elevate into an organizing principle. The belief about the special bond pertaining between natural scientists was a crucial resource in building Pugwash (Kraft and Sachse 2020). While there was indeed much that natural scientists had in common through their training and approach to tackling problems, a certain mystique was involved in assumptions about natural scientists' supposed access to unmediated and demonstrable truths about the material world and their capacity to retain an objective standpoint transcending political loyalties and bloc allegiances. This narrative could become a 'performative' element in the encounters at Pugwash conferences, and had powerful effects in cementing a sense of community and inspiring loyalty to the project that these scientists were embarked upon. At the same time, the durability of Pugwash encouraged a view amongst state actors that its conferences were worthwhile, even useful. Here, Moscow 1960 was decisive. This marked the point at which the dissenting origins of Pugwash in the west—whilst remaining a proud part of its heritage—gave way to a new era in which its western founders actively sought to find ways of working with governments in order to realize particular goals, even if their willingness to cooperate was accompanied by wariness. Tempering their scepticism about political establishments, these scientists—Rotblat, Powell, Rabinowitch—channeled their intellect and energies into finding new ways to realize the aims of the Pugwash organization. The thrust of their approach lay in developing a novel mode of techno-political communication conducted in informal and private settings, and which formed the basis to the developing role of the organization in Track II diplomacy. Here, its scientists capitalized on and added a further dimension to the transnational traditions of the exchange and flow of ideas, information and research findings long characteristic of international scientific networks.

Returning to the question as to the relevance of Pugwash to contemporary concept of Science Diplomacy, the analysis in this book makes clear the limitations inherent in drawing comparisons between the various modes of informal diplomacy as conceived and developed by Pugwash scientists and contemporary formulations of 'Science Diplomacy.'[1] Whilst they may share a view of the potential of science as a means to bridge international divides—for example, as set out in the case of Pugwash in the 1958 Vienna Declaration—and are centrally concerned with the intersection between science, diplomacy, politics and science policy making (nuclear and otherwise) they differ fundamentally in terms of the role of government actors at this intersection. Pugwash was critical of the superpowers for their development of nuclear arms and challenged the Cold War logic that was driving the nuclear arms race. Pugwash both asserted its independence (as a distinctive NGO-like organization) and eschewed formal ties with government, even as its scientists' devised ways and means to work with them. As and when Pugwash scientists worked with governments, it saw itself as an independent interlocutor developing a role in conflict moderation, working for disarmament, and as a broker for peace. By contrast, the programmatic thrust of the contemporary concept of Science Diplomacy posits a government-led 'soft power' model in which science is harnessed to foreign policy goals (Nye 1990, 2004). Here, science and scientists are brought into ever closer alignment with the state, a move which, in effect, places both in the service of the state as a means by which the state can realize specific goals.

Theories of the international dimensions of science can help illuminate this core difference. For example, Aant Elzinga and Catharina Landström's theorization of internationalism in science distinguishes between two broad models that differ with respect to the position or role of state actors in international initiatives (Elzinga and Landström 1996, 3–20). Briefly, Elzinga and Landström draw a fundamental distinction between scientist-led ('autoletic') and government-led ('heteroletic') forms of internationalism. The Pugwash project, including its activities in the diplomatic realm, can be seen as autoletic, whereas the primary influence of state actors in contemporary Science Diplomacy aligns with the heteroletic model. That is to say, each represents a wholly different approach to the use of science and to the role of scientists at the intersection between science and politics and within the realm of international diplomacy. The Pugwash project predated the contemporary concept of Science Diplomacy by half a century; both can be seen as responses to the very different geopolitical contours of the 1960s and the early decades of the twenty

---

[1] For a summary of this concept and the relevant literature, see the Introduction to this volume.

first century respectively. Naomi Oreskes has emphasized the need to place scientists in context to understand their actions and work (Oreskes 2014). The diverse and evolving roles of scientists at the intersection between science, politics, science (nuclear and otherwise) policy-making and diplomacy during the Cold War and beyond can only be understood by taking into account the political, economic, social and cultural contexts in which scientists operate and the specific constellation of forces acting upon them and on science: this is what shapes the use of both as resources for different actors, and for diverse ends. All of this underpins and explains the historical complexity of this intersection. The impact on society of what takes place at this intersection renders imperative the task of gaining historical understanding of the shifting relationships, dynamics and array of actors that defined this space—during and beyond the Cold War. Recently, the concept of Science Diplomacy has provided a powerful spur to scholarship in this direction, one that is illuminating the importance of science and scientists within the realm of international relations, and enriching the historiography of what Sönke Kunkel has called "the evolving entanglements between science, foreign policy, and international relations" (Kunkel 2021).

Three main arguments about Pugwash emerge from the analysis presented in this book. Each argument involves a paradox, testimony to the many contradictions at the heart of the PCSWA in this phase of its history.

Firstly, the innovative mode of techno-political communication developed within Pugwash enabled the leadership to reconcile its contradictory aims of working both with and against the grain of establishment politics and policy in East and West. This was evident at the Moscow 1960 conference. Earlier conferences had caught the attention of establishment scientists and political elites in the US sufficiently for representatives of the East Coast intellectual elite to attend the Moscow conference. What was experienced by participants there as a revelation was the sense that natural scientists could talk to each other across the political divides in conversations that took place privately and in an informal setting. This breakthrough was less about reaching agreement on disarmament—on the contrary, there were major differences across the table—and much more about the fact that they were talking and, indeed, developing a distinctive style of Cold War dialogue. Both because of the caliber and prominence of the scientists involved and because of the quality of the communication that took place, this conference changed opinions in government circles in the US and also in the UK about the Pugwash organization and its scientists. The Pugwash leadership too felt a breakthrough had been achieved in creating a new rapport with western scientific and political elites. Whilst Soviet scientists close to the Kremlin were regulars at the PCSWA, the organization had struggled to attract to its conferences senior western scientists close to the White House. But in late 1960, a change in the White House was in the offing and at Moscow Pugwash succeeded in drawing into its orbit senior scientific advisors close to the incoming Kennedy administration. The resulting encounters and private exchanges between influential circles in East and West constituted a new mode of unofficial or 'back-channel' diplomacy.

By now, Pugwash veterans within the Continuing Committee were already accustomed to off-the-record informal modes of working: this was how the Committee functioned and how the organization operated internally. But in Moscow the leadership was rolling out this modus operandi into a wider circle of scientists, who were in turn open to being drawn into a realm of informal diplomacy. In this way, communication—exchanges, discussion, dialogue—made possible by Pugwash reached the ear of government. As Eugene Rabinowitch succinctly put it at Ronneby, this was "private diplomacy addressed confidentially to the ruling elites."[2] Informal diplomacy did not replace official channels of negotiation. Indeed, some of those present in Moscow, for example Vladimir Arzumanyan and Richard Leghorn, were also active within official diplomatic channels in the disarmament talks in Geneva. But the additional possibilities offered by back-channel and informal contacts increasingly complemented the formal routes of official diplomacy.

The eighth conference in Stowe provided an early test of the viability of this strategy. For all the Cold War grandstanding in Vermont, it proved possible to continue the conversation about disarmament. Stowe thus consolidated the practices and forms of communication established in Moscow. It showed too that the Pugwash project could withstand serious political tensions between the US and USSR—indeed, arguably, part of its value lay in serving as a forum for confrontation between East and West precisely at such moments of crisis.

In accounting for the emerging practices of techno-political communication, the book pays particular attention to the expanded and enhanced discussions of specialist topics within the Working Groups at the conferences. These groups became productive sites of research into disarmament, and the reports arising from them were a valuable resource within and beyond Pugwash including, on occasion, within official East–West diplomatic channels. Working Groups also provided additional sites of transnational exchange and informal diplomacy. Beyond their importance in this regard, as the book has shown, Working Groups lent a more decentralized structure to the PCSWA and, on occasion, could serve as engines for 'change from below' in the Pugwash agenda.

At the same time, the conferences also provided the leadership a platform from which it branched out into unofficial diplomatic activities beyond the conference setting. Buoyed by the success of Moscow and having survived and learned from the difficult lessons in Stowe, the Continuing Committee experimented with a new targeted mode of back-channel diplomacy, drawing as they saw fit on the organization's resources—variously contacts, networks, know-how and experience on the international stage—to extend the diplomatic repertoire of the PCSWA. To this end, a new Executive Committee was established. If, in the 1960s, internally the organization existed in a state of flux with periods of instability, the leadership managed to maintain the reputation both of its conferences as a forum for cross bloc dialogue and sub-official diplomacy, and of its scientists as credible actors in this realm.

---

[2] Eugene Rabinowitch, The rationale of Pugwash. The seventeenth PCSWA, Ronneby, Sweden, September 1967. RTBT 5/2/1/17.

The second key argument of the book concerns political neutrality as both a guiding principle of Pugwash and as a set of practices within the organization. In all its forays into diplomatic territory, its identity as politically neutral was paramount. Externally, presenting the Pugwash organization to state actors as politically neutral was crucial to establishing its credibility, especially with Western governments. Demonstrating political neutrality was also important in bringing into the orbit of Pugwash scientific colleagues who had become part of the political establishment. Here, chapters one and two have shown the importance in the US case of the East Coast elite, and in the British case Solly Zuckerman, in changing perceptions of the PCSWA in Washington and in London respectively.

However, adhering to the principle of political neutrality was also a source of tension and turbulence within the organization. The difficulties caused by both the PSGES and by the Vietnam War mapped directly to this. The PSGES took a position on the German question that ran counter to the views of Washington and Bonn, and moreover, began both publishing these views and putting them forward in exchanges with external actors. In so doing, the group could be interpreted as jeopardizing the political neutrality of the PCSWA as whole. With regard to the Vietnam War, in refusing to criticize the US for any of its actions in Vietnam, senior figures within the western leadership of Pugwash were acting to uphold the principle of political neutrality, not least to protect its relationship with the Johnson administration in the US. But neutrality in this case could convey the impression of a leadership rendered immobile in the face of one of the great contemporary challenges of world politics in the mid-1960s.

The political neutrality of the Pugwash Secretary General—throughout this period a post held by Joseph Rotblat—was also critical in terms of shaping views of the PCSWA amongst external audiences. At the same time, this was crucial to how the organization functioned internally. In Rotblat's hands, the neutrality of the office of the Secretary General was a means of assuring the balance of East–West interests within the Committee and within the organization more generally. The Secretary General played the leading role in planning and choreographing each conference, but during the conferences Rotblat remained somewhat apart from the cut and thrust of debate, steering clear of anything that might be seen to erode the impartiality of the office. In this way, as Secretary General, he was better placed to mediate between colleagues at times of internal conflict and/or steer conversations towards compromise and restraint. As the book makes clear, Rotblat was also a driving force behind the organization's role in targeted back-channel interventions, where his reputation for impartiality and even-handedness was a strategic asset that carried weight with actors beyond the PCSWA. In short, Rotblat came to occupy a nodal position within Pugwash whilst serving also, externally, as gatekeeper to the organization. He was also an arch defender of its ways of working and values: on more than one occasion he fell out with colleagues who, in his view, had fallen short of the internal Pugwash code of utmost discretion and confidentiality.[3] Indeed, such was Rotblat's influence

---

[3] For example, in April 1963 Rotblat was furious with Bertrand Russell for what he viewed to be an inaccurate account of the Pugwash intervention in the Cuban crisis in Russell's recent book

within Pugwash in this period that it almost seemed as if the PCSWA was synonymous with him, which would be a misreading of both the man and the organization. Nevertheless, it is clear that Rotblat's Pugwash colleagues often looked to him as source of stability and leadership. In 1998, reflecting on his dissenting past, including as he put it, having in the mid-1950s been a 'whistleblower' about the dangers that nuclear weapons tests posed to human health, Rotblat remarked that "swimming against the current is not only strenuous but also dangerous" (Rotblat 1998, xvii–xxviii). Within the PCSWA, Rotblat and his colleagues had found through their fostering of communication between scientists another way to swim against the current as they sought to challenge the Cold War status quo and work towards disarmament and for peace.

For all the difficulties between the PCSWA and Washington, American influence within Pugwash at the international level remained strong throughout the 1960s, maintained by seasoned veterans of the organization including Rabinowitch, Bentley Glass, Harrison Brown, Bernard T. Feld and Geneva-based Martin Kaplan. All were routinely in touch with each other and with Joseph Rotblat by phone, letter, and in person—a circuitry of regular contact and sometimes friendships. This sustained a transatlantic axis vital to the organization on the international Cold War stage. When Rotblat finally stepped down in 1973, his successors in the post of Secretary General were Americans: Bernard T. Feld (1973–1976) followed by Martin M. Kaplan (1976–1988). That this important post remained the province of western scientists is a telling indication of the continuing need to advertise the western credentials of the PCSWA to Western governments and perhaps especially to Washington, with which relations remained uneasy. If at the same time this sent to the West a signal about the political neutrality of the organization, this was a signal that the Soviets—within Pugwash and in the Kremlin—could, seemingly, live with.

In shedding new light on the evolving culture within the PCSWA, the book reveals another paradox of Pugwash, namely, that for all the emphasis placed on developing and maintaining its much-vaunted informal modus operandi, the Pugwash organization was in practice deeply hierarchical. This is the third main argument of the book. At the apex of the PCSWA stood the Continuing Committee, including the Secretary General, and from 1963 onwards the Executive Committee. The leadership emphasized that the informal style of working conferred flexibility vital to the ability of Pugwash to react speedily to political crises. This may have been the case, but internally and in practice it accorded the Continuing Committee enormous discretionary powers and a means for retaining control over the development of the international organization and its constituent parts. An early indication of this concerned the national groups, the autonomy of which was, as Rotblat put it, contingent on their acting in ways that were consistent with "the chief criteria of the organization."[4] The book has cast fresh light on how internal tensions and divisions were handled by the

---

Unarmed Victory. Rotblat felt this had put him in an "embarrassing situation" that could "harm" his position in Pugwash and damage the organization. Rotblat to Russell, 13 April 1963. RTBT 5/1/2/6.

[4] Joseph Rotblat, 'Memo on future activities and organization,' 1962. RTBT 5/3/1/2.

Pugwash leadership using informal means and mechanisms, exemplified in the case of the Vietnam War and the PSGES.

Within the organization, the 1960s was marked by periodic but serious turbulence. The international PCSWA was profoundly affected by the wider geopolitical context in which it operated, and the mid-1960s Cold War brought new challenges. The book makes clear how the development of Pugwash at this point was shaped by the emergence of superpower détente, a landscape of disarmament transformed by various arms control treaties but still characterized by nuclear proliferation, by the political crisis in Central Europe (European security), and by the Vietnam War. A shift had taken place in which a wide range of different effects of the superpower rivalry were making themselves felt in international politics as opposed to the nuclear arms race between them per se. In this changed context, the unique asset of Pugwash— its scientific and technical expertise—appeared to be losing relevance to the most pressing international problems of the day. Moreover, the advent of groups such as SADS and SIPRI put an end to the Pugwash monopoly as an international forum for informal dialogue about disarmament. Meanwhile, for all the rhetoric about building links with and addressing the problems of the countries of the Global South, and for all Eugene Rabinowitch's efforts to this end, Pugwash made little headway in this direction.[5] By the late 1960s, this situation was drawing increasingly sharp criticism from within (Lavekare 1977).[6] Wherever it turned, the Pugwash leadership faced difficulty.

These geopolitical and institutional factors seeded the crisis that engulfed Pugwash during 1967–1968, as the organization lost momentum, influence and status, whilst the leadership was also struggling with internal divisions over its priorities and its agenda. But this crisis had another deep-seated dimension. For the founding cohort of Pugwash scientists, the generation of physicists born in the opening decades of the twentieth century who had been present at the onset of the atomic age, who had witnessed the nuclear arms race and the Cold War but who had dissented from the logic underpinning both, the mid-late 1960s brought a crisis of identity and purpose. This double dilemma—personal and about Pugwash—reflected the erosion of the consensus that had hitherto underpinned the PCSWA. The mid-1960s phase of the

---

[5] As early as the Moscow conference in 1960, Rabinowitch had stressed the position and needs of the "under-developed" world, calling for this region to be taken out of the superpower "power contest." This 'security' aspect was emphasized again by Rabinowitch in his paper at Ronneby, where he also called for the formation of a Study Group to consider the problems of the "new nations" and the expanding policy arena focused on Development. Rabinowitch, Eugene. Creation of a suitable climate for disarmament. Sixth Pugwash conference, Moscow 1960, *Proceedings*, pp. 688–708, especially pp. 694–697; The rationale of Pugwash. Seventeenth PCSWA, Ronneby, Sweden, September 1967. RTBT 5/2/1/17.

[6] After long-running discussions in the late 1960s and mounting pressure from scientists from the Global South, led by the Indian physicist Ashok Parthasarathi and the Ghanaian Frank Torto, the decision was taken at the 1970 Conference held in Fontana, US, to create a Pugwash Study Group on Development which came into existence in April 1971. Minutes of the 33rd meeting of the Continuing Committee, 10–16 September 1970, Fontana, US, and also of the 34th meeting, Frascati, April 1971. RTBT 5/3/1/2 (7).

Cold War was very different to that in which Pugwash had been founded and perceptions of the nuclear threat were shifting. In 1957, Hiroshima had felt very close and the threat of thermonuclear war imminent. Then there had been a strong sense in some parts of the international physics community that their profession, because of its involvement in the atomic bomb, carried special responsibility for the nuclear danger, a view captured by J. Robert Oppenheimer in his oft-cited observation after the first nuclear bomb test (Trinity) in the Alamogordo desert near Los Alamos in April 1945, that physicists had "known sin" (Cassidy 2005; Pais 2006; Thorpe 2006). These were the building blocks of the consensus that had driven the Pugwash project and bound its scientists together. In a period characterized by growing superpower détente and amid the changing dynamics of a global, multipolar Cold War, Pugwash scientists were searching for a consensus on which the organization could in the future be based.

The response to the crisis reasserted the primacy of nuclear disarmament on the Pugwash agenda and reasserted the authority of the Continuing Committee. In face of external challenges and internal turbulence, the Committee re-stated the core mission of the organization: Pugwash would focus on science and technology and function as a resource to which state actors could turn when scientific and technical expertise was called for. Organizational changes were made to facilitate this strategy. Notably, these included the discontinuance of the PSGES, which curtailed engagement with the German question and European security, and radically reduced the participation in Pugwash of lawyers, economists and social scientists. The simultaneous creation of the Pugwash symposia created a means to widen the scope of topics dealt with by the organization in a way that facilitated in depth discussion whilst setting limits on this. In addition, the Pugwash agenda began to place more emphasis on promoting international scientific and technical cooperation, especially with the countries of the Global South. In short, the Continuing Committee plotted a future path that played to its founding and distinctive strengths as an organization of natural scientists. The extent of its engagement with political issues was viewed through the prism of science and technology, reaffirming the primacy of natural science expertise to the identity and role of the PCSWA as it moved into the 1970s Cold War.

References

Cassidy, David C. 2005. *J. Robert Oppenheimer and the American Century*. New York: Pi Press.
Elzinga, Aant and Landström, Catharina. 1996. Introduction. Modes of internationalism. In *Internationalism and Science.*, ed. Aant Elzinga, and Catharina Landström. London: Taylor Graham.
Kraft, Alison, and Sachse, Carola. 2020. *Science, (anti-)communism and diplomacy: The Pugwash Conferences on Science and World Affairs in the early Cold War*. Leiden: Brill.
Kunkel, Sönke. 2021. Science diplomacy in the twentieth century: Introduction. *Journal of Contemporary History* 56 (3): 473–484.
Lavakare, Parabhakar. 1977. The Pugwash Movement in international affairs. *Foreign Affairs Reports* 26 (4): 84–98.

Nye, Joseph S., Jr. 1990. Soft power. *Foreign Policy* 80 (Fall): 153–172.

Nye, Joseph S., Jr. 2004. *Soft power: The means to success in world politics*. New York: Public Affairs.

Oreskes, Naomi. 2014. Introduction. Science in the origins of the Cold War. In *Science and technology in the global Cold War*, ed. Naomi Oreskes, and John Krige. Cambridge, MA: MIT Press.

Pais, Abraham. 2006. *J. Robert Oppenheimer: A life*. Oxford: Oxford University Press.

Rotblat, Joseph. 1998. A social conscience for the nuclear age. In *Hiroshima's shadow*, ed. Kai Bird, and Lawrence Lifschultz. Connecticut: Pamphleteer Press.

Thorpe, Charles. 2006. *Oppenheimer: The tragic intellect*. Chicago: University of Chicago Press.

Printed in the United States
by Baker & Taylor Publisher Services